The Anti-Nuclear Handbook

The Anti-Nuclear
HANDBOOK

Text by Stephen Croall
Illustrations by Kaianders

Pantheon Books
New York

Text Copyright © 1978 by Stephen Croall
Illustrations Copyright © 1978 by Kaianders

All rights reserved under International and Pan-American Copyright Conventions. Published in the United States by Pantheon Books, a division of Random House, Inc., New York, and simultaneously in Canada by Random House of Canada Limited, Toronto. Originally published in Great Britain as *Nuclear Power for Beginners* by Writers and Readers Publishing Cooperative.

Library of Congress Cataloging in Publication Data

Croall, Stephen.
The anti-nuclear handbook.

Bibliography: pp. 133-34.
1.   Atomic power—Caricatures and cartoons.
2.   Atomic power-plants—Caricatures and cartoons.
I.   Kaianders.   II.   Title.
TK9146.C68   1979   614.8′39   78-73645
ISBN 0-394-73781-4

Manufactured in the United States of America

24689753

A QUICK WORD

SC & KS

The Anti-Nuclear Handbook

THIS IS YOUR CAPTAIN SPEAKING...

GOOD AFTERNOON AND WELCOME ABOARD ATOMIC AIRWAYS.

WE ONLY HAVE FUEL FOR THE EARLY STAGES OF THIS FLIGHT, BUT WE THINK WE CAN SOLVE THAT EN ROUTE... THE OTHER SAFETY PROBLEMS MAY ALSO BE SORTED OUT IN TIME... OUR TECHNICAL BOYS HAVE BEEN WORKING ON THEM FOR 30 YEARS, AND THEY'LL LET US KNOW IF THEY COME UP WITH ANY ANSWERS...

AS FOR SABOTAGE... WELL, IT'S ALWAYS A POSSIBILITY... AFTER ALL WE LIVE IN DANGEROUS TIMES... BUT YOU'LL HAVE NOTICED THAT YOU'RE ALL HANDCUFFED TO YOUR SEATS AND GAGGED... SO THE CHANCES ARE PRETTY REMOTE... FINALLY I'M SURE YOU'D LIKE TO KNOW OUR DESTINATION...
SO WOULD WE...
ENJOY THE TRIP...

HERE COMES
CHAPTER 1.

In which the atom is split by mistake, some 200,000 Japanese are killed on purpose, Sam & Ivan play in the sandpit and the Indians call everyone's bluff...

Scientist are funny people...

Give them an **atom**, tell them it's the smallest thing around and what do they do? Split it in two, of course. Just to rub it in, they'll split the very heart of the atom — the **nucleus**!

The scientists spent the first three decades of the 20th century probing the nature of the atom...

NIELS BOHR OF DENMARK **ERNEST RUTHERFORD** OF BRITAIN

Unlocking the atom, though, proved a tough nut until a Briton called **James Chadwick** discovered a tiny particle in 1932 — the **neutron**...

By bombarding atoms with neutrons scientists found they could turn one chemical into another...and this became a favorite pastime in the labs of the 1930s...

Among the most skillful neutron-tossers was the Italian Enrico Fermi...

**One day they tried bombarding the nucleus of a uranium atom...
and it disappeared! Something was up...**

OTTO HAHN.

In 1939 an Austrian physicist, Lise Meitner, rescued the scientific community from going crazy by suggesting the uranium atom had *split in two*...a process she called
FISSION

Ms Meitner's decisive role in this historic achievement was rewarded with a Nobel Prize...for her ex-partner Otto Hahn!

A fission reaction

A neutron hits the nucleus of a uranium atom...

...the nucleus absorbs the neutron but becomes unstable...

...and splits in two, releasing some energy + 2 or 3 **new neutrons**...

The release of fresh neutrons opens up the prospect of a CHAIN REACTION...

One neutron fissions one uranium atom, releasing *two* neutrons that can produce two more fissions, releasing *four* neutrons etc etc...

The fissions occur every 50-billionth of a second so a lot of energy is released very rapidly...

7

MAKING THE A-BOMB...

US President Roosevelt was told that if the Western Allies didn't stick that chain reaction inside a weapon quickly the mad A. Hitler would get the bomb first...so in 1941 he gave the go-ahead for

The Manhattan Project

The cream of US, British, Canadian and French scientists got together to make a nuclear fission bomb...

Fermi, the Italian whiz kid, played a key role.
Some have called him "the greatest physicist of the century". Others remember his stock response when colleagues muttered about the A-bomb posing a threat to humanity:

DON'T BOTHER ME WITH YOUR CONSCIENTIOUS SCRUPLES. AFTER ALL, THE THING'S SUPERB PHYSICS!

$E = mc^2$
$^{235}_{92}U$ $^{239}_{94}Pu$
$\alpha + \beta + \gamma + n$

Fermi & Co. reached their first goal in December 1942. A clumsy uranium reactor built in a converted squash court under a Chicago sports stadium worked at the first try...

The energy it produced would not have lit a flashlight ...but it marked Planet Earth's entry into **THE ATOMIC AGE...**

AWFULLY SECRET

I STILL DON'T UNDERSTAND WHY FRANKIE WANTED TO BOMB MANHATTAN...

By now US scientists had turned up another fissionable element that was even better than uranium for making bombs... PLUTONIUM...one of the deadliest substances ever created by Man.

F.D.R

ALL NUCLEAR REACTORS BURN URANIUM AND PRODUCE PLUTONIUM!

Over the next couple of years the bomb materials — plutonium and *enriched* uranium — slowly mounted up ...the project was so hush-hush that Harry Truman didn't even know about it when he took over the presidency on Roosevelt's death in 1945...

A New York journalist who nosed up evidence that nuclear energy could make an almighty big blast discovered later that his articles had been classified top secret — after publication.

DAILY SNOOPER
BIG BANG CAN WIN WAR!
URANIUM

9

By the time they were ready to test, in July 1945, some of the scientists were having serious doubts about the morality of a nuclear fission bomb. But they were outnumbered. The test, at **Alamagordo** in the New Mexico desert was a shattering success and terrified some of the bomb's creators. Asked to describe what he'd seen, one physicist replied: "I can't tell you, but . . .

DON'T EXPECT TO DIE A NATURAL DEATH!

CLAP CLAP

The Manhattan Mob lost no time in demonstrating to the world how greatly they'd improved the human race's capacity to destroy itself...on August 6, 1945, a uranium bomb nicknamed *Little Boy* was dropped on the Japanese city of Hiroshima. It worked and the war was as good as over... But the boys back home were anxious to try out their plutonium bomb as well... So three days later *Fat Man* was dropped on the city of Nagasaki... That one also worked.

The two raids killed some 200,000 people immediately. Another 100,000 died from the after-effects of radiation over the ensuing decades. Today, victims of **Little Boy** and **Fat Man** are still dying in Japan at the rate of 2,000 a year... ,

World War II came to an abrupt end...
The development of nuclear weapons did not...
The US Navy was sulking because
the US Army had gotten all the
glory for outstanding services
to mankind at Hiroshima and
Nagasaki...

The marines
needed
to
restore
their
manhood...

YAAAAH!
I WANNA GO
BOOM TOO!

BOOM

OK. KID

So in the summer of 1946
they arranged for 42,000
spectators to witness two bangs
at a Pacific atoll
called Bikini...

BOOM
BOOM...

THE BIKINI ATOLL

Absent from this display of
heroics were the local islanders...
they'd been secretly removed a few months earlier...
and could not return until 22 years later...by which time the
vegetation had become unproductive and the menu was
radioactive coconut crab...

PEACE
FOR
PRODUCTION
PROGRESS
PROFIT.

MEANWHILE,
Back in the States, the
politicians and businessmen
were getting excited. Uncle
Sam was the only one **in the
whole wide world** with an
A-bomb! Time to spread some
Pax Americana around the
globe by gentle persuasion...
And not a moment too soon,
what with the weirdo commies
fanning out in Eastern
Europe and the **Yellow Peril
going red in China**...

JUST TO BE ON THE SAFE SIDE
Sam passed the **McMahon Act.**
This neat little law meant
that anyone who thought
the Manhattan Project
had been a joint effort
was out of his mind.
The thing had been built
in Sam's back yard
so it was only fair that
all nuclear information
be labeled:

The **British,** of course, were absolutely furious...They still hadn't got over the way the bloody Yanks had taken all the bloody credit for winning the bloody war...**Now Sam was showing them the nuclear door!** They stormed home in a huff to build their own bangs...

The **Canadians** were
so upset they
decided not to make
any bombs at all...

13

**So the stage was set
for the United States of the World...**
or so it might have seemed to some...
Then **Ivan's Gang** went and spoiled
everything by
testing a fission
bomb of their own in
September 1949.
It was six times
bigger than
Little Boy...

THE RUSSIANS
ARE COMING!

Now it was common knowledge that no goddam
Russky could get the better of Sam except by cheating...

Scapegoats were soon found —
a couple called Rosenberg
who were accused of slipping
Ivan **the secret formula...**
In the thrill of a trial and
execution most people
overlooked
the fact that the
'secret formula' had long
been in general circulation
in the scientific community...and that
Ivan had been fully engaged in nuclear research since 1940...

The Soviet mushroom launched
the Arms Race...also known as
the Sandpit Syndrome...

MINE IZ
BIGGER!
HE HE...

HO HO, MINE
CAN GO FARTHER!
HAHAHA!

BANG!

WHO SLIPPED THEM THE SECRET?

In October 1952 Britain's military scientists stopped sulking long
enough to blow the fish out of the water outside Australia...

14

London newspapers started tossing around phrases like **The Big Three**... but no one was really fooled... fission bombs were kid stuff by now and the two superpowers were working on superbombs combining fission with **fusion**...

DAMN FINE! RECAPTCHA INDIA FIRST THING MONDAY.

ISN'T HE CUTE!

IMPERIAL RAG

KENYA: MAU-MAU WOG TERROR!

VIET GOOKS CRUSH FROGS IN INDO-CHINA

FAT JOE STALIN DIES AT LAST?

A-BOMB OBSOLETE!

BIG THREE!

BRONZE FOR BRITAIN.

2

1

3

Fusion is slamming together **hydrogen nuclei**✳ to form **helium** and create a **thermo-nuclear chain reaction**...

HI - WANNA BLOW A FUSE?

SMACK!

ENERGY

RADIATION

HEAT

JUST MARRIED

WOW

Whereas the fission bomb could destroy cities the H-bomb could destory whole regions.

NOW, THAT'S WHAT I CALL PROGRESS!

Sam gave it a ground run in November 1952, wiping a Pacific island off the map with a blast 500 times bigger than Hiroshima...

✳ One nucleus, two nuclei. Latin form. Ask any of the old Romans.

15

Ivan dropped an even bigger one in August 1953…which Sam topped in March 1954… but made himself a bit unpopular by raining the radioactive fallout on a couple of hundred islanders 150 kilometers away…and on the Japanese fishing vessel *Lucky Dragon*…

One of the crewmen died.

Thus the first victim of the **H-bomb** was Japanese, as were the first victims of the **A-bombs**… So it's not surprising that US repairmen often have to be flown in when nuclear power stations go wrong in Japan today…the natives are a bit sensitive about radiation levels…

16

MEANWHILE,

Sam was at war again — disguised as a United Nations soldier — trying to win Korea for the Free World... Leading the charge was World War II hero **General MacArthur**, who was so keen on liberating the place that he asked for atomic bombs...

President Truman didn't think the idea sounded too bad...but he was talked out of it by his pals at the UN...and Mac lost his job...

GODDAMMIT, HARRY! I PROMISED THE BOYS THEY'D BE HOME FOR CHRISTMAS!

KOREA

Anxious as ever to play with the big boys, Britain let off an H-bomb in 1957...and over the next decade France and China tested atomic and hydrogen bombs to join the **Fission & Fusion Freaks...**

3-2 TO THE CAPITALISTS

Why stop at the H-bomb? What about a weapon capable of blowing the world to bits? Scientists have considered this tempting project... on the principle that the pursuit of truth knows no boundaries...but ruefully concluded that the **Doomsday Bomb** is a technical impossibility...

SORRY TO LET YOU ALL DOWN!

EDWARD TELLER FATHER OF THE H-BOMB

17

By way of consolation Sam's military brass gave us:

First tested in 1977, this delightful bit of hardware has one major advantage — **it does little damage to buildings**...It just spews out neutrons (*remember the neutrons?*) and kills all life within range...any survivors are thus spared the bother of building up a new production apparatus...

But we're getting ahead of the story...

The Arms Race accelerated into the 1970s with Sam and Ivan battling for the lead, the Chinese a good third and the French and British gamely hanging on...

Suddenly out of nowhere came the Indians!

18

THERE WAS AN IMPORTANT DIFFERENCE BETWEEN INDIA'S PLUTONIUM BOMB TEST, IN MAY 1974, AND ALL PREVIOUS NUCLEAR TESTS...

WE WEREN'T SUPPOSED TO BE ABLE TO MAKE BANGS... WE ONLY HAD A SMALL CANADIAN RESEARCH REACTOR, IMPORTED FOR PEACEFUL PURPOSES...

It was all terribly embarrassing.

For years we'd been assured by
      the scientists
      the military
      the industrialists
      the politicians

PEACEFUL NUCLEAR POWER CAN **NOT** BE USED TO MAKE NASTY NUCLEAR BOMBS! AND THAT'S FINAL!

YEAH, I AGREE!
HEAR, HEAR!
RIGHT!
IMPOSSIBLE!
CAN'T BE DONE.
WELL MAYBE...

PEACEFUL NUCLEAR
TEST.

AGGRESSIVE NUCLEAR
TEST.

Had someone blown their cover 10 years earlier
things might have turned out differently...

But by the end of Indian Bomb Year more than 150 of those harmless
little nuclear power plants (*nukes*) were sprouting up around the
globe...

The 'exclusive' Nuclear Arms Club now looks like
getting a dozen new members during the 1980s...

THIS IS
CHAPTER 2.

In which «peaceful» nukes spread around the globe, the oil giants become energy giants, the Human Factor gets into his stride and accidents tend to happen…

Originally the heat generated in the fission process was considered a useless by-product...

But by the late 1940s a lot of people realized reactors were more than just a way of producing plutonium...

Some people got carried away...

To the European industrialists nuclear power sounded like a godsend. They were running short of cheap fuel just when a Second Industrial Revolution was in the offing. The fact that **nukes can only generate electricity** and that electricity has very limited industrial uses wasn't mentioned much...

The military had set the wheels in motion, laying on investment, research and development of nuclear power ...the capitalists just had to climb aboard...

Over at Sam's place private industry wasn't into nukes...there was plenty of cheap oil and the prospect of lots of coal and natural gas...

WHO NEEDS NUKES

1¢ A GALLON

It quickly changed its mind when the **federal government** threatened to go into the business...

THOSE WHITE-HOUSE COMMIES HAVE GOT TO BE STOPPED!

Canada was the first country to concentrate on civil nuclear programs...and, in 1952, the first to play host to a major accident. A research reactor at Chalk River exploded but no one was killed so no one got excited...

BOOM

WE'RE PROUD TO ANNOUNCE THE FIRST **INDOOR** NUCLEAR EXPLOSION.

Sam & Ivan boasted sizable reactors by 1954...but **Britain** (Calder Hall) and **France** (Marcoule) claimed the first **commercial** nukes in 1956, generating electricity to the public...

In the same year the **International Atomic Energy Agency [IAEA]** was set up in Vienna. It aimed to **promote the spread of nuclear power** and **prevent the spread of nuclear weapons materials.** Like teaching a child to walk then ordering it not to...

Sam took a short-cut...in 1957 he called home the nuclear-powered submarine **Nautilus,** stole its reactor and set up the first US nuclear power station at Philadelphia...

Ivan built *his* first civil nuke at Troitsk in Siberia...a breakthrough for Leninism??

Also in 1957
the nature of nuclear power was
demonstrated by a fire at Britain's
Windscale plant...a radioactive cloud floated
out and two million liters of milk were poured
away because of the contamination risk...

By this time, several
countries *that did not
have nuclear arms* were
charging head-first into
the nuclear age...West
Germany...Canada...Italy
...Japan...Sweden...

All the nuclear states had one thing in common — they'd taken the plunge without giving the general public a chance to express an opinion about nuclear power. The scientists had consulted the military, the military had consulted the politicians, who consulted the businessmen, who had a word with the scientists, and round and round they went...

25

In fact, not until 30 years after nukes made their appearance were the people of any country asked in any way whatsoever if they wanted this extremely dangerous technology...

Today, so much state and private capital has been poured into nuclear power expansion that a critical public is faced with a massive wall of interlocking economic interests and misleading propaganda...

BACK TO HISTORY. The 1960s opened with a bang — at the Idaho Falls SL-1 nuclear plant in the US. Three repairmen were killed...

Their bodies were so contaminated by radiation that they couldn't be buried for 20 days...and then only in lead-lined coffins sealed in lead-lined tombs...

The accident was blamed on human error...as were the accidents at Chalk River and Windscale...**and most of the mishaps that were to plague every major reactor design before the end of the decade!**

The worst reactor accident to date may or may not have been the one in 1963 involving Sam's nuclear submarine **Thresher**. It disappeared on a deep dive with 129 people aboard. No one knows for sure if it fell victim to a reactor breakdown...

Three years later came the confrontation everyone had been dreading...**the Human Factor met the Breeder Reactor!**

WHO ARE YOU?

BREEDER

I'M THE GREAT WHITE HOPE OF THE NUCLEAR INDUSTRY. TELL YOU WHY LATER.

They crossed paths at the Enrico Fermi experimental plant outside Detroit. Someone forgot to fasten a bit of metal in the reactor core; it caused part of the fuel to melt and orders went out to prepare for the evacuation of the city...

SOME CRAZY GUY WANTS US TO EMPTY DETROIT! I TOLD HIM TO TAKE A PILL AND GET SOME SLEEP.

A disaster was averted...but only just. Had it occurred, claiming hundreds or thousands of lives, the development of nuclear power would probably have halted...or at least fast breeder development...

DON'T COUNT ON IT.

Instead, the second half of the 1960s brought an upswing in sales. Nuclear electricity was now looking profitable so Sam's heavies decided it was time to muscle in...

MOVE OVER SMALL FRY.

GENERAL ELECTRIC

WEST ING HOUSE

The giant US firms had got off to an early start so they were well placed to seize control of the Western reactor market...and they still dominate today...

I MAKE MY OWN NUKES AND SHIP THEM TO MY FRIENDS

ANATOMY OF AN OIL CRISIS...

In 1973 the Arab oil-producing countries jacked up basic prices by 400%...and the 'black gold' that had fueled the rapid industrial build-up of the West suddenly became 50% more expensive to the consumer...

They knew they were emptying the oilfields too fast...there were nationalization rumors all over the Middle East...and that they'd have to branch out to stay in business...but that required immense investment...and the low price of oil had meant it was unprofitable to move into coal, gas or nukes...

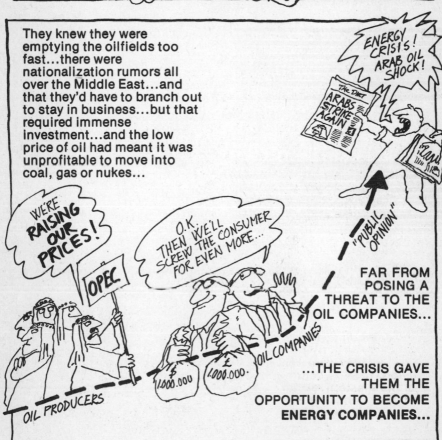

FAR FROM POSING A THREAT TO THE OIL COMPANIES...

...THE CRISIS GAVE THEM THE OPPORTUNITY TO BECOME **ENERGY COMPANIES**...

UPSHOT OF THE OIL CRISIS...

NUCLEAR POWER LOOKED THE BEST BET...

§ It promised to yield greater profits than coal

§ There was no union militancy in the nuclear industry ...while coalminers showed growing reluctance to die for the bosses

§ Uranium reserves were in politically 'safe' places like the US, Canada, Australia and...er...southern Africa

Although Sam had a firm grip on the reactor-building business by the 1970s he was not without rivals...**The British** for example had come up with a nuke they called the AGR. It was such a mess that it's no longer mentioned in polite company...

The Swiss also had a crack at making one...and wisely stuck it underground, at Lucens. After two years the core blew up and the reactor had to be scrapped...

Sweden's first attempt was even more embarrassing...it couldn't be made to work at all! The plant, at Marviken, was refitted to run on oil...

By contrast, *Canada's* CANDU reactor proved a money-maker. Exactly why became clear when India startled everyone with its bomb test in 1974 — the CANDU is a very efficient producer of plutonium...

31

But the nukes that outsold them all were Sam's **LIGHT WATER REACTORS** [LWRs]…Governments and power companies were in such a hurry to buy that they didn't worry too much about minor details like rotten safety records…

The 1973 'oil crisis' had opened a lot of doors to the nuclear salesmen. By the mid-1970s there were 19 countries in the nuclear fold…among them two from the Third World…

THE HUMAN FACTOR

meanwhile, was having a gay old time…he fitted one reactor back-to-front, fitted pumps upside down in another, forgot a welding rig inside a third, lost track of uranium shipments and plutonium stocks, and shut down nukes left, right and center…

THE HUMAN FACTOR's

favorite toy was the LWR...

In 1975 he almost wrecked a couple at Browns Ferry in Alabama. A 7-hour fire knocked out emergency cooling in one reactor, closed down another and damaged hundreds of cables. The cause? A *candle* carelessly handled by an electrician...

OH MY GOD, NOT HIM AGAIN...

I ALMOST LOST THEM THE TENNESSEE VALLEY...

Despite the antics of HF the nuclear power industry predicted a dramatic worldwide expansion...with nukes in 40 or 50 nations by the end of the century meeting 50% of global electricity 'needs'...

...and the Fast Breeder just waiting to solve all our energy problems in the next century...

YUM, YUM GIMME! GIMME!

BUT SOMETHING WENT SOUR... In the second half of the 1970s a lot of people woke up to the dangers of nuclear power...and threw the whole future of the industry into doubt...

CLOSED FOR NON-PROFIT REASONS

$1,000,000

33

THE INDIANS HAD DEMONSTRATED THAT THERE WAS NO SUCH THING AS A **PEACEFUL ATOM**... BUT THAT DIDN'T WORRY THE INDUSTRY.

THE CLAIM THAT NUKES WERE RELIABLE AND SAFE RANG INCREASINGLY HOLLOW WITH EVERY BREAKDOWN AND DISASTER... THAT **DID** WORRY THE INDUSTRY. A BIT...

BUT NUKES BEGAN TO LOOK LIKE A **BAD ECONOMIC INVESTMENT**, AND THAT **REALLY** UPSET THE INDUSTRY!

$ 1,000.00

WHEN YOU'RE TRYING TO MONOPOLIZE THE WORLD ENERGY SUPPLIES, HIGH RISKS AND LOW PROFITS ARE THE LAST THINGS YOU NEED!

In the US where half the world's nuclear trade goes on reactor salesmen suddenly found doors being slammed in their faces...and orders being postponed or cancelled. It was all too much for the major oil companies...

WHY DO YOU REFUSE TO PUBLISH SEPARATE ACCOUNTS FOR YOUR REACTOR BUSINESS?

WELL... ER... I'M JUST NATURALLY MODEST...

BP

WESTINGHOUSE

Shell, Gulf & Co. started shedding their reactor interests...only Westinghouse of the big corporations claimed to be making a profit in the nuclear field...

PROF.IT

In Japan and West Europe too...nuclear programs are far less ambitious today than a few years ago...some Western states have been scared off nukes altogether...while in Sweden a government lost office after 44 years to an opponent who pledged to scrap the nuke...

THE PROMISE WAS BROKEN, OF COURSE...

IN THE BEST PARLIAMENTARY TRADITION.

As for Britain...home of the first 'commercial' nukes...the domestic industry hasn't sold a reactor since 1970 and looks in danger of collapse...

SPECIAL! BARGAIN! BUY BRITISH!

Everywhere, the evolution of a military-run industry into a civil one has brought enormous problems...technical, economic, political and moral...now the industry is on the defensive, hardly able to cope with its exposure to public scrutiny and the open marketplace...

AT MY PLACE, WE HAVE NO SUCH PROBLEMS, SO IT'S FULL STEAM AHEAD.

35

OOPS! THESE DAYS YOU CAN'T EVEN SPLIT AN ATOM WITHOUT FINDING A CRITIC INSIDE...

ATOMS UNITE!

OPPOSITION TO THE NUKE
has spread around the world faster than the technology itself...

The objections to nuclear power are numerous... some can equally well be leveled at the other big-scale, centralized energy technologies like coal and oil...

YOU MEAN THERE'S NO GREAT DIFFERENCE BETWEEN NUKES, OIL AND COAL?

EVER TRIED BUILDING A COAL BOMB? OR AN OIL MISSILE?

HEY, THAT'S AN *IDEA*...

BUT WHAT DISTINGUISHES NUCLEAR POWER FROM ALL ELSE IS IN A WORD:

RADIATION

DON'T MISS CHAPTER 3.

In which the China Syndrome is explained, reactor cores are cooled or caught, Rasmussen puts our minds at rest and the Third World is ripped off once again...

WHAT IS NUCLEAR RADIATION, AND WHERE DOES IT COME FROM?

It's bursts of energy — sub-atomic fast-flying particles or high-energy electromagnetic waves — released when the nuclei of unstable (radioactive) atoms try to reach a stable state…

OK. BUT WHAT'S THIS GOT TO DO WITH ME?

BRATATATA…

Once inside your body these little beasts attack your cells… a direct hit on a gene can make your biological master-plan go haywire…in a word —**cancer!**

OOPS! NASTY. IS THERE MUCH OF THIS AROUND?

There used to be unstable atoms all over the place but by the time life evolved on our planet nearly all of them had calmed down…now nukes and A-bombs and things are turning the clock back millions of years…recreating lots of these immature, unstable young monsters…

HEY! STOP THAT!

SO MUCH FOR PROGRESS THEN…

NO NUKES

RADIATION MAY BE A NATURAL PROPERTY OF MATTER, BUT IT'S BY NO MEANS A PART OF MAN'S NATURAL HABITAT. IN FACT, **IT'S DEADLY DANGEROUS!** ESPECIALLY FOR TINY CREATURES LIKE ME!

ALPHA RAYS
BETA RAYS
GAMMA RAYS
NEUTRON RAYS
POSITRON RAYS

RADIATION IS WHAT MAKES NUCLEAR POWER SO RISKY!

Huge amounts of radioactive material are produced in the operation of nukes…the 'ashes' from the nuclear 'burning' of uranium fuel…and only a small leak can cause a catastrophe…

I'LL TAKE THAT RISK.

1,000,000. 1,000,000. 100,000.

HEHE… LET'S BREAK HIS MOLECULES APART!

YEAH! AND LET'S GIVE HIM CANCER!

THE REALLY NASTY THING ABOUT RADIATION IS THAT YOU CAN'T SEE IT OR FEEL IT OR HEAR IT OR SMELL IT OR TASTE IT OR TELL IT TO GO AWAY…

WE RADIOACTIVISTS JUST SNEAK IN AND MAKE OURSELVES AT HOME IN YOUR BODY, DESTROYING TISSUE, NERVES AND GENES, AND YOU KNOW NOTHING ABOUT IT UNTIL IT'S TOO LATE…

THE NUKE IS BOTTLED-UP RADIATION!

Were there any guarantee that it could be *kept* bottled up…after leaving the reactor as well…people wouldn't be making such a fuss…but there is no guarantee…

WERE IT NOT FOR THE RADIATION, NUCLEAR POWER SIMPLY WOULDN'T WORK!

A TYPICAL NUCLEAR POWER REACTOR...

The heat from the fission process boils the water to steam

The steam spins the turbine which turns the generator thereby creating electricity

The water is pumped back to the reactor vessel to repeat the cycle

The sea or river provides coolant to condense the steam back to water

The discharged coolant is several degrees warmer

ORDINARY POWER STATIONS WORK IN THE SAME WAY EXCEPT THAT THE WATER IS HEATED BY BURNING OIL OR COAL...

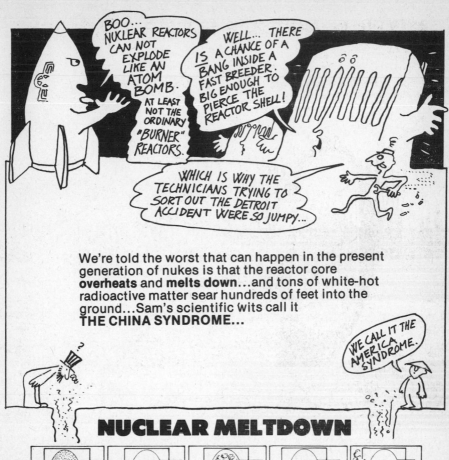

We're told the worst that can happen in the present generation of nukes is that the reactor core **overheats** and **melts down**...and tons of white-hot radioactive matter sear hundreds of feet into the ground...Sam's scientific wits call it **THE CHINA SYNDROME**...

NUCLEAR MELTDOWN

A main coolant pipe ruptures and the core temperature rises.

The fuel itself melts and the core fills with hot radioactive gases.

Steam explosions fling metal parts through the dome walls...while the molten uranium burns down into the Earth.

The emergency cooling system also fails and the fuel rods melt.

The molten fuel burns through the pressure vessel and containment dome.

The most dangerous forms of radioactivity likely to be unleashed in a meltdown are the gaseous **fission products.** The exact consequences of such an accident — how many people would die how quickly — could only be established by a **test run**…

A key barrier between a meltdown and a public health disaster is the ECCS — the emergency core cooling system. There's a lot of evidence that it's less than totally reliable. Special hearings in 1973 turned up two dozen safety researchers **from the US Atomic Energy Commission itself** with misgivings about the ECCS…

Of course the nuclear industry is **fully confident** that the ECCS is a safe back-up…so confident that it's developing 'core catchers' in case the ECCS fails. These are designed to disperse and cool the molten debris. The French are installing one beneath their massive **Super-Phénix** breeder at Malville…

IMAGE THAT THE WORST COMES TO THE WORST... Your local reactor is devastated by a major accident and a radioactive cloud drifts into your neighborhood...

Those invisible particles aren't good for you so your best bet is to leave. They can settle into your food, drink, building materials or anything else that's lying around...**worst of all, though, is breathing them in...**

43

> HOW MANY OF US WOULD BE KILLED IN A REACTOR DISASTER? ...AND WHAT ARE THE CHANCES OF IT HAPPENING?

For years lots of busy little men have been putting such questions to their slide-rules and computers. Depending on which report you trust it seems that between a dozen and tens of thousands of victims could be expected...

> NOT WHERE I COME FROM!

A test run would be useful to the statisticians...but it might be difficult to keep out of the newspapers...

A major accident contaminating a large area apparently took place in the Soviet Union in 1958. Ivan hasn't let on what happened. It could have been a reactor breakdown but most signs point to an explosion of radioactive waste buried underground...

CLOSED FOR IDEOLOGICAL REASONS.

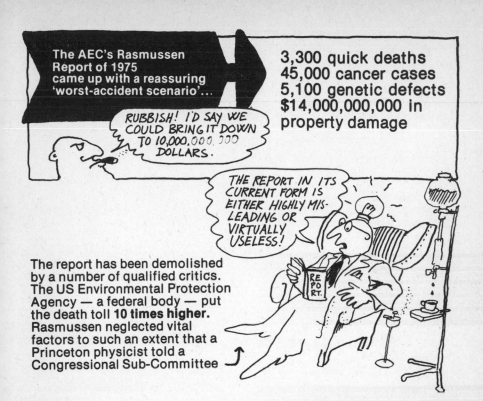

The AEC's Rasmussen Report of 1975 came up with a reassuring 'worst-accident scenario'...

3,300 quick deaths
45,000 cancer cases
5,100 genetic defects
$14,000,000,000 in property damage

RUBBISH! I'D SAY WE COULD BRING IT DOWN TO 10,000,000,000 DOLLARS.

THE REPORT IN ITS CURRENT FORM IS EITHER HIGHLY MIS-LEADING OR VIRTUALLY USELESS!

The report has been demolished by a number of qualified critics. The US Environmental Protection Agency — a federal body — put the death toll **10 times higher.** Rasmussen neglected vital factors to such an extent that a Princeton physicist told a Congressional Sub-Committee

Rasmussen's models were two plants in the US...where the law says no more than 380,000 people may live within 25 kilometers of a nuke. But elsewhere the situation can be very different...for instance, Sweden's Barsebäck plant is smack in the middle of Scandinavia's most densely-populated area...

By the way

IN 1975 THE U.S. ATOMIC ENERGY COMMISSION FISSIONED INTO TWO ORGANISATIONS.

Atomic **E**nergy **C**ommission

Nuclear **R**egulatory **C**ommission

Energy **R**esearch and **D**evelopment **A**dministration

Over a million Swedes and Danes live within 25 kilometers of Barsebäck and two million live within a 40 kilometer radius. The Danes have no nukes themselves and are naturally thrilled at the Swedes' willingness to involve them in the risks...

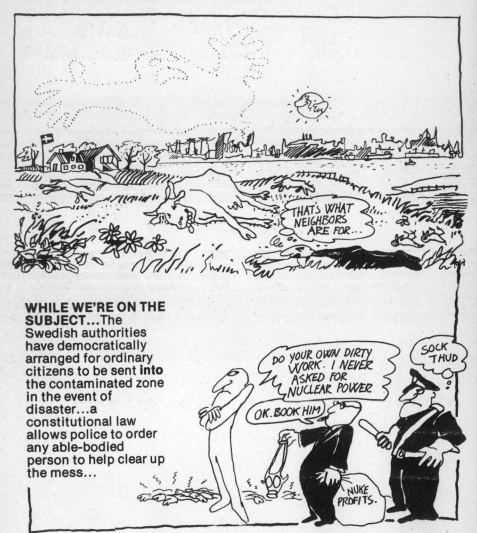

WHILE WE'RE ON THE SUBJECT... The Swedish authorities have democratically arranged for ordinary citizens to be sent **into** the contaminated zone in the event of disaster...a constitutional law allows police to order any able-bodied person to help clear up the mess...

Rasmussen decided the chances of a disastrous accident occurring were extremely remote...

AND THE FOLLOWING DAY ONE DID?

THERE WAS JUST AS MUCH CHANCE THAT A METEORITE WOULD HIT A U.S. CITY, HE SAID.

No, but the **Brown's Ferry** accident was just the kind he was more or less ruling out...

Moral: Rattlesnakes are harmless unless they bite.

RAZZLE DAZZLE

A later study by the Union of Concerned Scientists in the US put the probability of a nuclear plant catastrophe **100 times higher** than Rasmussen...

WHICH MEANS THAT BY THE END OF THE CENTURY WE MIGHT BE HAVING SERIOUS ACCIDENTS EVERY SIX YEARS...

47

The nuclear power enthusiasts specialize in minimizing risks using two main arguments...

THERE ARE OTHER RISKS IN SOCIETY JUST AS BAD!

THE ODDS ARE HEAVILY AGAINST A REACTOR DISASTER!

1. One risk does not excuse another...
2. If an accident can have **catastrophic results** the chances of its happening are of secondary interest (unless there is no chance whatsoever)...

IT'S A BIT LIKE SAYING THAT I'M SAFER THAN SHE IS

WHICH IS TRUE, IF YOU JUST COMPARE THE ODDS AGAINST US FALLING... OOOPS!

SLIP

THE BIG DIFFERENCE IS IN THE **CONSE-QUENCES** OF THE FALL!

SPLAT!

WHY DO THE NUCLEAR AUTHORITIES KEEP INSISTING THAT THEIR NUKES ARE SAFE?

MY DEAR FELLOW, NO NUKES - NO NUCLEAR AUTHORITIES...

MOST REACTOR SAFETY PROGRAMS IN THE WEST RELY ON THE FINDINGS OF THE AEC/NRC...ALTHOUGH IT HAS A LONG AND IMPRESSIVE RECORD OF MISLEADING THE PUBLIC ON THE SUBJECT...AND ON JUST ABOUT EVERY OTHER NUCLEAR MATTER...

After the special ECCS hearings a Congressional Committee said the AEC had **'developed a serious credibility gap...by suppressing unwelcome evidence of danger and by demoting or firing researchers who have pushed their findings too seriously...'**

The New York Times studied 11 years of AEC documents and concluded that the agency **'repeatedly sought to suppress studies by its own scientists that found nuclear reactors were more dangerous than officially acknowledged or that raised questions about reactor safety devices...'**

Leo Goodman, a union energy expert sometimes called the godfather of the anti-nuclear struggle in the US, put it more bluntly: **'AEC was dishonest and tricky from the word go...Watergate was nothing new to me...'**

49

FAR FROM BEING SAFE nukes have been besieged by problems from the outset…as pointed out in a 1973 AEC report which was suppressed until the environment group Friends of the Earth got hold of a copy…

The British among others have flatly rejected Sam's claims that his LWRs are safe…a program to buy 32 of them was abandoned after the British government's Chief Scientific Advisor decided they could rupture without warning…

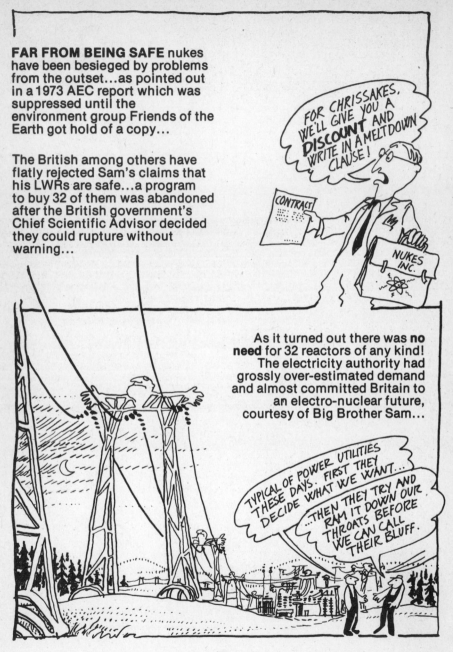

As it turned out there was **no need** for 32 reactors of any kind! The electricity authority had grossly over-estimated demand and almost committed Britain to an electro-nuclear future, courtesy of Big Brother Sam…

Operating nukes on home ground is a risky enough exercise ...exporting them to non-industrial countries is a recipe for calamity...

IN FACT, THERE MAY NOT EVEN BE AN ORIGINAL MODEL!

WOULD YOU MIND EX-PLAINING YOUR-SELF, YOUNG MAN.

Reactors are a bit more than a source of heat in a bucket of water...they're very complicated...and when fully built the **exported** nuke may be very different from the **original** model in the supplier state.

The countries exporting to the Third World — France, West Germany, Canada and the US — mainly build big ones for the domestic market...nukes generating 1,000 Megawatts (MWe) of electricity or more...

MEGAWHAT?

MEGAWATT! THAT MEANS ONE MILLION WATTS OF ELECTRICITY!

BUT WE OF THE 'LESS-DEVELOPED COUNTRIES' DON'T NEED GIGANTIC POWER PLANTS.

FOR SALE.

O.K. SO YOU GET A SITUATION LIKE THIS...

The true horror story

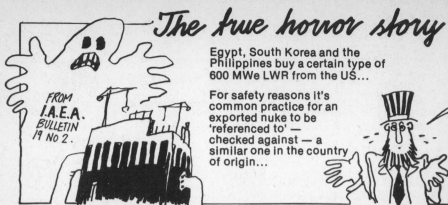

Egypt, South Korea and the Philippines buy a certain type of 600 MWe LWR from the US...

For safety reasons it's common practice for an exported nuke to be 'referenced to' — checked against — a similar one in the country of origin...

FROM I.A.E.A. BULLETIN 19 No 2.

There's another problem...a reactor working like a dream on the banks of Lake Ontario may not be at all happy in Western Africa. Take the case of two West German LWRs in Iran...

MEIN GOTT! FRITZ, I THINK I'M GOING TO THROW UP.

They're being built in a relatively high seismic area but they're referenced to a plant in a low seismic area of West Germany

SO VOT?

SO WE MUST MAKE A LOT OF DESIGN CHANGES.

YES! FROM THE FOUNDATIONS...

...TO THE REACTOR COMPONENTS...

...AND THESE WILL NOT BE SUBJECT TO REVIEW BY THE WEST GERMAN SAFETY AUTHORITIES!

IAEA

of the nuclear rip-off:

BUT BECAUSE OF THE HOME DEMAND FOR **BIG** REACTORS, WE'VE NO CORRESPONDING MODEL FOR THEM TO REFER TO.

One being built in **Puerto Rico** was called off before completion...had it been finished it would have undergone systematic safety inspection and...

A NUMBER OF ALTERATIONS WOULD UNDOUBTEDLY HAVE BEEN MADE.

PROJECT ABANDONED

IAEA

So instead the Egyptians, Koreans and Filipinos have to 'reference' their nuke to a similar one in Yugoslavia...which had been referenced to an earlier plant in Brazil...which in turn had been referenced to the abandoned Puerto Rican plant!

PUERTO RICO

BRAZIL

YUGOSLAVIA

WELL, AT LEAST THAT LEAVES US PLENTY OF SCOPE FOR IMPROVISATION.

Important changes and modifications made in domestic nukes in the exporting states are not necessarily included in the nukes sold abroad...

BUCKS
BUCKS
ORIGINAL MODEL

THE MAIN VICTIMS, OF COURSE, ARE THE IMPORTING COUNTRIES LACKING SAFETY REVIEW BODIES OF THEIR OWN.

WHICH MEANS JUST ABOUT THE ENTIRE THIRD WORLD, MAN!

Just around the corner is the next generation...

THE FAST BREEDER REACTORS

LET'S BREED, BABY!

They're a bit different from today's burner reactors...

All nukes produce plutonium but only the breeders are designed to run on it...they generate more fissionable material than they consume... and in theory they're eventually self-supporting...

INCREDIBLE! THIS MEANS WE CAN ESCAPE THE URANIUM CRISIS.

I MIGHT HAVE GUESSED.

THERE'S JUST ONE SNAG...

It takes time for a breeder to produce enough plutonium to refuel another breeder...*not just years but decades!*

This 'doubling time' is put at 40 to 60 years for the French experimental plant in Marcoule — one of only a few operating breeders in the world today — but they're hoping for 30 to 40 years at the Super-Phenix in Malville...

IF THEY EVER GET IT FINISHED

COUNTRIES WITH COMMERCIAL NUKES
OPERATING, UNDER CONSTRUCTION OR PLANNED

1975 - Argentina	1980 - Austria
Belgium	Brazil
Britain	Finland
Bulgaria	Hungary
Canada	South Korea
Czechoslovakia	Mexico
France	Taiwan
East Germany	Yugoslavia
West Germany	
India	1985 - Iran
Italy	Philippines
Japan	Poland
Netherlands	Rumania
Pakistan	South Africa
Soviet Union	
Spain	1990 - Cuba
Sweden	Egypt
Switzerland	Israel
United States	Thailand
	Turkey

For the plutonium collectors among you:

A LUMP, HALF THE SIZE OF A TENNIS BALL, CAN WRECK A SMALL CITY.

If nuclear power grows as fast as planned commercial reactors will be churning out enough plutonium each year by 1990 to make 45,000 atom bombs...

BY THAT TIME I'LL JUST BE GETTING INTO MY STRIDE.

BIG BREEDER PROGRAMS WILL MEAN A NEW AGE IN WHICH ONE OF THE WORLD'S MOST VITAL RAW MATERIALS IS HIGHLY POISONOUS!

A FULL-SIZED BREEDER COSTS SOME $ WHICH MEANS YOU CAN'T EASILY REVERSE A DECISION TO BUILD ONE ONCE YOU'VE STARTED.

YUM YUM

PLUTONIUM ECONOMY

ON TOP OF EVERYTHING, THEY LOOK PRETTY UNSAFE!

GRRR GRR

Breeders are 'cooled' by liquid metal...usually sodium... which reacts explosively with water or air. So a leaking pipe can be a bit of a problem...

DA, THAT'S HOW MY BREEDER AT SHEVENKO BLEW IT IN 1974.

OOPS... FORGET EVERYTHING I SAID...

57

STAND BY FOR
CHAPTER 4.

In which a cycle is not a cycle, radioactive waste is not disposed of, the spread of nuclear technology is not stopped and societies can not protect themselves from nuclear sabotage except by converting to a police state...

BUT NUCLEAR POWER IS NOT JUST A LOAD OF REACTORS.

IT'S A WHOLE CHAIN OF COMPLICATED LINKS KNOWN AS:

THE URANIUM MINE

Uranium ore is mined and concentrated into *yellow-cake* — a mixture of two uranium types [*isotopes*] called U-235 & U-238. Only U-235 can be fissioned to give energy but it makes up less than 1% of the uranium...

THE BREEDER REACTOR

Besides running on plutonium it converts U-238 into *more* plutonium... which can be used after reprocessing to fuel more breeders...

NUCLEAR ARMS PRODUCTION

About 10 kilos of plutonium is enough to make a crude nuclear device...

PLUTONIUM STORAGE

Plutonium is kept for possible re-use as breeder fuel...or put aside for military use...

UCLEAR FUEL CYCLE.

THE ENRICHMENT PLANT
By various techniques the U-235 content of the uranium is raised to 2-4%...

FUEL CONVERSION PLANT
The enriched uranium is converted into oxide pellets which are packed into fuel rods ready for the reactor...

THE REPROCESSING PLANT.
The spent fuel is broken down by acids...the unburnt uranium and plutonium are recovered...and the waste remains...and remains...and remains...

THE NUCLEAR REACTOR
The fuel fissions in the reactor core to generate heat for electricity production...and plutonium and radioactive 'fission product' waste...

WASTE STORAGE
Highly radioactive [high-level] waste is so toxic that it must be isolated from all life *for centuries or millenia!* Today it's stored while experts try to figure out how to get rid of it...

DANGER

61

AND THIS IS WHAT THE
LONDON FINANCIAL TIMES
WOULD HAVE US BELIEVE...

NO UNPLEASANT
LEFT-OVERS LIKE
RADIOACTIVE WASTE OR
HOARDS OF PLUTONIUM...

IT JUST GOES ROUND AND
ROUND AND ROUND...

In reality it's more like a game of hide and seek played out in a minefield. All along the way there are hazards to public health...but it's the worker in the nuclear industry who's the most directly threatened... especially in the mines, in reactor repairs and waste 'management'...

The left-overs after uranium is extracted from ore are called *tailings* and they're radioactive ... which came as a bit of a shock to people in the southern states who had found them lying about in sand-like heaps and decided they'd make fine foundation and building material...

BEFORE...

A SPECK OF INHALED PLUTONIUM CAN GIVE YOU CANCER...

...AFTER?

SAFETY AT ENRICHMENT AND FUEL FABRICATION PLANTS IS NOT A BIG PROBLEM because the uranium fuel is still in a fairly tame state...*but the large-scale use of plutonium as fuel would change the picture completely...*

64

The *Russians* and *Chinese* have their own, in Siberia and Lanchow. The European Common Market countries have ganged up to build commercial plants in *Britain* (Capenhurst), *France* (Tricastin) and the *Netherlands* (Almelo)...

ALSO, OF COURSE, URANIUM ENRICHMENT CAN PROVIDE THE RAW MATERIAL FOR A-BOMBS...AS THE PEOPLE OF HIROSHIMA LEARNED IN 1945...SO DESPITE THE COST ENRICHMENT FACILITIES ARE QUITE A POPULAR ITEM IN THE NUCLEAR CATALOGUE...

Also in the race are *Canada* (James Bay), *Japan* (Tokai Mura) and *South Africa* (Valindaba). A test plant began operating at Valindaba in 1975...which may explain why South Africa was in a position to conduct a nuclear bomb test in 1977...

65

AS FOR THE REACTOR...

The nuclear industry is always issuing publicity photos like this to show us how nice and safe it is to work in a nuke...

For some reason photos of repairmen in action in a real emergency don't find their way into the PR brochures... and very rarely into the newspapers...

MAYBE PHOTOGRAPHERS AREN'T TOO KEEN ON RISKING THEIR LIVES...

The Swiss nuclear power industry once ran out of top-flight welders! All 700 who were qualified for the job collected the maximum radiation dose when repairing a reactor at Beznau...radiation was so intense that each welder could work for only two minutes...

On a breakdown at the Indian Point reactor in New York state *2,000 repairmen* were used in the same way...

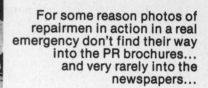

HI MAN. HOW ARE THINGS?

GREAT! JUST GOT TWO MINUTES WORK!

EMPLOYMENT OFFICE

AND SO TO
CHAPTER 5.

In which resources are husbanded, energy is renewable, decision-making is shared and a Soft Energy Future seems the only rational course...

SAYING 'NO' TO NUKES
MEANS SAYING 'YES' TO SOMETHING ELSE!

Little would be won if it meant saying 'yes' to ferocious consumption of coal, oil & gas...their risks are well-known...

* As a workplace the coal mine is as hazardous as the uranium mine...

* The death rate for oil industry divers in the North Sea is notoriously high...

* Tankers carrying frozen natural gas can explode like a small A-bomb...at sea or in harbor...

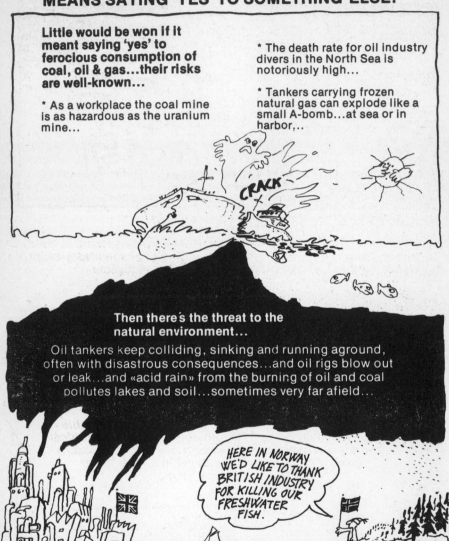

Then there's the threat to the natural environment...

Oil tankers keep colliding, sinking and running aground, often with disastrous consequences...and oil rigs blow out or leak...and «acid rain» from the burning of oil and coal pollutes lakes and soil...sometimes very far afield...

HERE IN NORWAY WE'D LIKE TO THANK BRITISH INDUSTRY FOR KILLING OUR FRESHWATER FISH.

Also, burning coal, gas & oil raises the carbon dioxide content in the Earth's atmosphere...heating it up...so continued use of fossil fuels at today's rapid rate can upset the world climate...

BUT COMPARING THE RISKS IS MISSING THE POINT.

Must we choose between cholera and the plague…

…or can we instead go for a clean bill of health?

Technological development and social development go hand in hand. Today's decisions in the energy field are going to shape lifestyles into the next century…and the effects could be hard to reverse…

HERE'S ANOTHER KIND OF CHOICE…

Either we move on down the *hard technology road* towards a *high-energy society* in a *nuclear age* based on a *plutonium economy*…

…or we take the *soft energy path* towards an *equitable society* based on *ecological principles* and an *economy geared to people's real needs*…

69

The official line in the industrialized countries is that demand for energy will continue to soar...and that around 1990 it'll outstrip supplies of fossil fuels...creating an ENERGY GAP that only the nuke can fill...

IN FACT, THEY SAY, WITHOUT NUCLEAR ELECTRICITY OUR LIVES WILL BE TURNED UPSIDE DOWN BY THE END OF THE CENTURY...

REACTORS AND JOBS...

US estimates show that big power stations produce fewer jobs per dollar than just about any other major investment! British estimates show that the electrical supply industry is the most capital-intensive of them all....

Amory Lovins:

66 It's the conservation, solar, environmental and related social programs that yield the most energy, jobs and monetary returns per dollar invested...
The huge, capital-intensive energy facilities often proposed to relieve unemployment not only make it worse by draining from the economy the capital that could make more jobs if invested almost anywhere else, but also worsen inflation by tying up billions of dollars non-productively for a decade. **99**

THAT'S HOW LONG IT TAKES TO BUILD A BIG POWER STATION THESE DAYS.

THE LONG AND SHORT OF IT...

Abandoning the high-energy, hard technology approach is going to create *short-term* difficulties for certain groups...for instance workers making nuclear components. In some countries the bosses have jumped at this opportunity to drive a wedge between the trade unions and the anti-nuclear power movement...

But this divide-and-rule tactic is not always a success...

THE ALTERNATIVE SOCIETY
GOES UNDER MANY NAMES...

And takes many shapes. No single model fits everywhere. In fact, it's not so much a question of models as a strategy for change at various times with specific goals in mind...

Tactics vary. Some people think the profiteers can be reformed...that the ruling élites can be talked out of their privileges...that national governments can control multinational monopoly capitalists. Some think the environment crisis can be dealt with by juggling prices and tariffs...

But the idea is growing that only a socialist blueprint for change can reverse today's destructive trends...

THE ALTERNATIVE GOALS...

§ Switch-over from fossil fuels to renewable energy forms...
§ Conservation of natural and human resources...
§ Global solidarity...
§ Meaningful production and work for all...
§ Grass-roots government...

Measures like these are known as *technical fixes*...and they're important because of the tremendous wastage from inefficient energy use. Some experts claim that savings of 40-50% are possible in today's industrial structures...and 30% in homes and transport...

Changing over to the naturally-distributed energy sources is ecologically essential. But they can be developed in different ways...

THE MALIGNANT WAY
To prop up the centralized and authoritarian structures of our existing industrial society...

FOR EXAMPLE THERE ARE PLANS AFOOT FOR:

A vast solar power station in orbit beaming energy to Earth by microwave...

THE BENEVOLENT WAY
To promote the decentralization of political and economic power, redistribution of wealth and the liberation of the individual...

A gigantic battery of 300,000 wind turbines on the American Great Plains...

SO IT'S NOT **JUST** A QUESTION OF HARD OR SOFT...HIGH OR LOW...BIG OR SMALL...

ANY ENERGY STRATEGY MUST ALSO BE JUDGED BY ITS POLITICAL AND ECONOMIC IMPLICATIONS...

IT DEPENDS HOW YOU LOOK AT IT...

Top: THE OFFICIAL PICTURE...
BLUE SKIES & UNSPOILED NATURE...
AND IT'S SO HARMLESS YOU CAN
LIVE RIGHT NEXT DOOR...

Middle: THE REACTOR AS IT MIGHT
BE SEEN BY THE OPPONENT OF
NUCLEAR POWER WHO JOINS A
DEMONSTRATION...

Right: THE HOUSEHOLD REACTOR
(MID-21st CENTURY) AFTER BEING
TAKEN OUTSIDE BY THE REPAIRMAN...

OVER AT THE RE-PROCESSING PLANT

There are lots of plutonium and radio-active fission products being handled ...so almost everything has to be operated by remote control from behind thick shielding...

The French plant at La Hague was the only one by 1978 capable of treating LWR oxide fuel...a lot of countries were lining up ... and as the order-book swelled the management exposed the workers to increasing doses of radiation. In the face of union protests more and more unorganized, short-time labor was employed...often with only a couple of days' training!

TROUBLE IS, CLEAN-UP AND REPAIR WORK SOMETIMES REQUIRES HUMAN BEINGS...

La Hague workers told union investigators...

OFTEN NO CONTAMINATION CHECKS WERE MADE AT THE EXITS...

I'VE SEEN WORKERS GO HOME EVEN THOUGH THE DETECTION METERS WENT BERSERK AS THEY APPROACHED...

I SAW A FELLOW WITH BREATHING TROUBLE LIFT HIS GAS-MASK IN A RADIO-ACTIVE ZONE TO GET A BIT OF "FRESH AIR"...

The workers at La Hague had to strike for months before the French nuclear industry would do anything about their working conditions...

... RT TO GET THINGS STA... REALLY MESSY

A reprocessing accident can be at least as dangerous to the public as a reactor accident...yet the reprocessing plant lacks most of the reactor's safety barriers...

Radioactivity could escape into the environment if the witches' brew leaked out of the acid pond...or the coolant failed and the waste boiled...or the ventilation packed up and there was a chemical explosion...

REPROCESSING IS SUCH A DIFFICULT AND RISKY BUSINESS THAT IT HAS MORE OR LESS BEEN ABANDONED IN THE U.S....WITH GIANT LOSSES. GETTY OIL, GENERAL ELECTRIC, GULF AND SHELL HAVE ALL BURNED THEIR FINGERS...

Some people haven't given up, though. New plants for reprocessing oxide fuels are planned or being built in Britain (Windscale), West Germany (Karlsruhe), India (Tarapur) and Japan (Tokai Mura)...and La Hague aims to expand...

Finding a final resting place for high-level nuclear waste is a problem that has baffled scientists ever since the stuff started piling up in the 1940s...

Interim storage is a neat way of passing the buck...but liquid radioactive waste is a sizzling peril. In the United States more than 400,000 gallons have seeped away from one storage facility at Hanford on the Columbia River...

There was also a big leak from Windscale's storage tanks in 1976...

The Windscale management reacted in the true spirit of the nuclear establishment, which has always had the welfare of the general public at heart...

How nuclear cover-up went wrong

TOP officials at British Nuclear Fuels Ltd decided from the start to cover up the facts about a radioactive leak from their Windscale plant because they feared repercussions at a critical moment in the site's development.

Even when they were urged by officials from the Government's Health and Safety Executive to make the facts public they still said nothing, although by staying silent they contravened the spirit of the Nuclear Installations Act.

ALL PROPOSALS FOR LONG-TERM STORAGE OR DISPOSAL OF HIGH-LEVEL WASTE FROM THE NUCLEAR POWER INDUSTRY LIE AT THE RESEARCH AND DEVELOPMENT STAGE...WHILE STOCKPILES GROW FASTER AND FASTER AROUND THE GLOBE...

There have been lots of *theoretical* solutions...

But in the words of the Nobel Prize-winning physicist Hannes Alfvén: *'If a problem is too difficult to solve one can not claim that it's solved by pointing to all the efforts made to solve it...'*

BECAUSE OF THE IMMENSE TIME IT TAKES FOR HIGH-LEVEL WASTE TO COOL DOWN AND BECOME HARMLESS NO METHOD CAN BE GUARANTEED COMPLETELY SAFE...

❝ *We must assume that these wastes will remain dangerous and will need to be isolated from the biosphere for hundreds of thousands of years. In considering arrangements for dealing safely with such wastes man is faced with time scales that transcend his experience...* ❞

— Flowers Report to British Parliament September 1976

Exhaustive research has failed to turn up anything but a couple of reasonably promising ways of solidifying the waste...and the bright idea that it would be best stored where it's retrievable in case things go wrong...

CAN'T YOU LEAVE THE SPENT FUEL INSIDE THE RODS INSTEAD OF REPROCESSING IT?

THERE'D BE FEWER PROBLEMS... BUT IT WOULD STILL REQUIRE ULTIMATE DISPOSAL.

AND IT WOULD MAKE ME REDUNDANT...

How long it takes for high-level waste to become docile depends on how long it takes for its ingredients to decay. Most of the *fission products* have calmed down after 800 years...but other radioactive elements called *actinides* survive longer, including traces of plutonium not stripped out of the fuel in reprocessing...

AFTER ONE HALF-LIFE HALF OF THE ORIGINAL MATERIAL REMAINS.

AFTER TWO HALF-LIVES, A QUARTER (HALF OF HALF) OF THE ORIGINAL MATERIAL REMAINS.

AFTER THREE HALF-LIVES, AN EIGHTH. AFTER FOUR, A SIXTEENTH. ETC.......

1 — ½ — ¼ — ⅛ — 1/16

ONE HALF-LIFE

TWO HALF-LIVES

TIME

The rate-of-decay of a waste product is measured in its **half-life** — the time it takes for half the amount of material to disintegrate. For example Caesium-137 has a half-life of 30 years...so reducing radioactivity a thousand-fold takes 300 years (ten half-lives). Species like Strontium 90 (**28 years**), Curium-244 (**18 years**) and Caesium-137 are more dangerous than long-life ones like Plutonium-239 (**24,400 years**) because their radioactivity is more concentrated...

THE PROBLEM IS THAT HEAVY, RADIOACTIVE ELEMENTS LIKE PLUTONIUM DECAY INTO OTHER RADIOACTIVE ELEMENTS, WHICH DECAY INTO OTHERS AND SO ON...AND DON'T FORGET THAT, APART FROM THEIR RADIATION, THESE ELEMENTS ARE HIGHLY POISONOUS!

High-level *waste is not the only problem*...there are *medium-activity* and *low-activity* left-overs as well...in much larger quantities...

Low-level waste is routinely discharged into the surroundings, buried in the ground or dumped at sea...

Most medium-level waste is toxic enough to require isolation for centuries. It can include anything from protective clothing to old bits of reactor...

DISUSED REACTOR- AND REPROCESSING PLANTS ARE ALSO A FORM OF WASTE... RADIOACTIVE RUINS THAT WILL STAND AS MONUMENTS TO... WELL... ER... SOMETHING...

CLICK

WHERE DOES THE BREEDER COME IN?

IT'S SUPPOSED TO *CLOSE* THE FUEL CYCLE...BY MAKING USE OF PLUTONIUM...WHICH MEANS A LOT OF REPROCESSING...AND THERE AIN'T MUCH AROUND THESE DAYS...

AND THE FRENCH ARE PLANNING BREEDERS SPECIALLY FOR ME

IRAQ BRAZIL

So it's not strange that the people going for breeders are also going for reprocessing... the Soviet Union, Britain, France, West Germany, Italy, India and Japan...

President Carter set the cat among the pigeons in 1977 by halting commercial reprocessing and plutonium recycling in the US. But his appeal for others to follow suit was met with suspicion...

HA! IF YOU ARE SERIOUS, WHY DON'T YOU CANCEL YOUR **CLINCH RIVER** BREEDER?

YOU JUST WANT TIME TO CATCH UP, N'EST-CE PAS?

ZU LATE! ME UND MEIN FREUND ARE TAKING OVER YOUR MARKETS, UNDERSTAND?

BREEDERS FOR SALE

*

The only way to close the fuel cycle is to shut your eyes and pretend there's no such thing as radioactive waste...

THE FUEL CYCLE STRETCHES ROUND THE WORLD...

I'M NOTHING IF NOT INTERNATIONAL!

Japan may buy uranium from South Africa, convert it in Britain for enrichment in the US, push it through a reactor at home, send the spent fuel to France for reprocessing and get back a load of plutonium...

PLUS A LOAD OF HIGH LEVEL WASTE.

THAT'S OK. WE'LL BURY IT IN SOUTH KOREA.

Many think the risk of accident, theft or sabotage is greatest during the transport of nuclear materials around the fuel cycle...

They're moved by road, rail, sea and air...and today's flow of traffic promises to swell to a flood over the next two decades...

ACCIDENTS HAPPEN!
SHIPS SINK!
TRUCKS CRASH...
SO DO TRAINS & PLANES!

PEOPLE GET CARELESS!
PEOPLE GET CORRUPTED!
PEOPLE GET BLACKMAILED!
THIEVES STEAL!

TERRORISTS
GO IN
SHOOTING!

HOLD ON... WE TAKE THE UTMOST PRECAUTIONS WITH NUCLEAR FREIGHT

SURE! UNTIL THEY GET TOO EXPENSIVE OR SLOW YOU DOWN.

85

THE NUCLEAR PARK SOLUTION...

The industry claims that the latest reactors are built to withstand a crashing plane. But not a Jumbo-size airliner. Nor a big bomb. As for reprocessing plants, they're almost totally unprotected from things falling out of the sky...

IN THE EVENT OF WAR...NUCLEAR POWER IS A LIABILITY...

Nukes are such obvious and potentially disastrous targets that the only sensible thing to do if hostilities break out is to *close them down*...so countries with most of their energy eggs in the nuclear basket will be fighting the war in the dark...

A TRIP DOWN PARANOIA GULCH...

What people usually mean by the 'wrong hands' are those belonging to guerilla groups, organized crime syndicates, stray madmen or leaders of the Third World...
as opposed to characters of
infinitely sound judgment
like Richard Nixon, Moshe
Dayan, NATO and Warsaw Pact
generals...

The name of the game is...
CLANDESTINE DIVERSION OF PLUTONIUM
or...
NUCLEAR PROLIFERATION AT GOVERNMENTAL
AND NON-GOVERNMENTAL LEVEL
or quite simply...
SPREADING THE BOMBS AROUND

Two years after the Indian Bang showed *Atoms for Peace* and *Atoms for War* to be Siamese Twins the record was officially put straight by an NRC Commissioner:

International action to control (the dangers of proliferation) associated with the civilian nuclear fuel cycle depends critically on the understanding of two facts. First, that nuclear weapons can be manufactured from reactor-grade plutonium. And second, that for any nation that has done its homework, separated plutonium can be suddenly appropriated from its storage place and inserted in warheads within days.

FOR SALE...

You don't need a big nuclear power program to make nuclear weapons under the counter...as the Indians demonstrated. A small research reactor...laboratory-scale reprocessing...and the whole thing was said to cost only $400,000...

A humble little 40MWe reactor gives enough plutonium a year for three Hiroshima-size bombs. The Stockholm International Peace Research Institute (SIPRI) says the components can be discreetly bought on the open market for $20 million...and small reprocessing and uranium enrichment units are also available to those anxious for bomb materials...

THE NEW SUPERSALESMEN...

"THE SECRET OF OUR SUCCESS IS THAT WE'RE UNPREJUDICED - WE'LL DEAL WITH ANYONE."

"JA, RACISTS, FASCISTS, MILITARY DICTATORS, RELIGIOUS FANATICS... YOU NAME IT."

"YOU CAN SAY THAT AGAIN."

"GOD BLESS THE FREE MARKET"

SOUTH AFRICA

LIBYA

France and West Germany won a place in every pacifist's heart by contracting to sell the direct route to nuclear bombs — missing links in the fuel cycle — to Pakistan and Brazil respectively.

Today, in the shadow of the Soviet-American nuclear arms race, local rivals are engaged in contests of their own. The Pakistanis are scrambling to catch up with the Indians...the Brazilians are vying with Argentina for nuclear leadership of Latin America...and both Egypt and Israel are already believed to have access to nuclear weapons.

NONE OF THESE NATIONS ARE A PARTY TO THE NON-PROLIFERA-TION TREATY [NPT] WHICH WAS SUPPOSED TO STOP THE SPREAD OF NUCLEAR ARMS...

"DOWN HERE IN SOUTH AFRICA WE NEED NO ENCOURAGE-MENT TO DEVELOP A NUCLEAR THREAT..."

WHO'S AFRAID OF THE **N.P.T.** ?

A lot of non-nuclear countries, for a start…They see it as a neat way for the big powers to hog the nuclear hardware. And they have some cause for resentment…The British, Americans and Russians have consistently refused to promise they won't use their nuclear arms against non-nuclear states — even the ones that sign the NPT!

HOWEVER… WE MIGHT BE PREPARED TO NEGOTIATE THE **NUMBER OF WARHEADS** WE'LL USE ON YOU…

DA!

OUR GRATITUDE KNOWS NO BOUNDS

NON-NUCLEAR STATES

If you read the small print it's very clear why many countries choose not to sign the NPT…*the treaty obliges signatories to help non-signatories with nuclear energy programs wherever possible!*

WASHINGTON?… WE DEMAND OUR RIGHTS UNDER THE **NPT.** SEND A COUPLE OF REACTORS, ONE OF THOSE REPROCESSING THINGS, AND A SIX-PACK OF URANIUM 235.

CASH ON DELIVERY?

THAT'S HOW INDIA GOT HOLD OF THAT CANADIAN REACTOR…IT JUST WENT AND ORDERED IT!

90

FROM ONE PAPER TIGER TO ANOTHER...

Even when it's signed the NPT is not much of a deterrent as membership can be cancelled on 3 months' notice...or simply ignored in a 'national emergency'...then you just clap a warhead onto the missile the NPT allows you to assemble and off you go...

Many experts doubt that legal safeguards are enforceable anyway... Treaties invariably have loopholes...nuclear inventories are quite easy to fiddle...bomb tests can be simulated on computers...and an inspector's task can be made virtually impossible on site...

So the nuclear 'haves' decided to lean on the 'have-nots' a bit more heavily...they formed THE LONDON CLUB...

The London Club claims it wants to stop the spread of sensitive technology and know-how to the 'wrong quarters'. At the same time, its members are competing with one another in a cut-throat multi-billion dollar export industry...

AND WE ALL KNOW THE UPSHOT OF CONFLICTS LIKE THIS.

WHO'S WILLING TO LOSE MARKETS JUST FOR THE SAKE OF WORLD PEACE?

so it's no surprise that they can't agree which countries to sell their goodies to under what circumstances...or rather which countries not to sell their goodies to...

GUESS WHO'S MAKING MOST OBJECTIONS?

LONDON CLUB'S A BUNCH OF SISSIES

Slightly more promising for the future of humanity was Carter's 1977 initiative to strangle reprocessing and breeders and his announcement that recipients of US nuclear exports would be required to place them under the IAEA safeguards...

Carter also got 40 nations to join in an International Fuel Cycle Evaluation (INFCE)...and although it implied a two-year pause in plutonium development around the world the French and West Germans were among them...

SEE THAT? I STILL CARRY SOME NUCLEAR MUSCLE.

MEANWHILE DOWN IN THE CELLAR:

Reactor-grade plutonium is not ideal for bomb-making because it's a bit jumpy...

THE CHAIN REACTION IS LIKELY TO OCCUR BEFORE THE DESIRED FORCE HAS BUILT UP.

SO THAT IT GOES OFF TOO EARLY, AND SAYS POP INSTEAD OF **BANG**.

ALL BUT THE MOST DESPERATE GOVERNMENTS WOULD AVOID RELYING ON SUCH INEFFICIENT BOMBS.

...but urban guerillas or criminal groups might not be choosy!

Authorities like the NRC say reactor-grade plutonium can be used by such outsiders *on the basis only of existing information in the open literature* to make crude but convincing bombs...of unpredictable yield but enough to wipe out a seat of government...

SIPRI SUMMED UP THE HAZARDS LIKE THIS...

66 Plutonium could be stolen and sold on a «black market», or it could be ransomed, or a group of criminals might steal plutonium for profit, or for use as a nuclear threat to deter police or otherwise further their activities. Or one of the more than 50 terrorist groups that are said to exist worldwide might see nuclear weapons as means of enhancing its capability to use, or threaten, violence. Or a revolutionary-minded political group within a country might aquire nuclear weapons to achieve its political objectives or to deter violence against it. And it should not be forgotten that the danger inherent in a crude nuclear device constructed by, for example, a terrorist group is not confined to a possible nuclear explosion. The contamination of a large area by high levels of plutonium would be an enormous threat in itself. 99

A London newspaper which received a 'credible' A-bomb design from a group of students asked some scientific and weapons experts to describe in concrete terms the amateur nuclear weapon threat...

THE TERRORIST A-bomb is about the size of a tea-chest, weighs around half a ton, could be conveniently carried in an inconspicuous vehicle such as a Transit van, and would be detonated remotely by radio.

If the yield of one kiloton was achieved its fireball would be about 70 metres in diameter, vapourising much of the area where it touched the ground. All exposed people within a radius of half a kilometre would be killed directly by thermal and other radiation, while devastation of structures by blast might extend over a radius of about 1 kilometre. In some meteorological conditions, such as a surface temperature inversion which is common in Britain at dusk and during the night, blast damage might extend over a radius of up to 3 kilometres.

Because this would be a surface burst the radioactive plume would be large and would contain a great deal of particulate material which, recondensed from the fireball, would contain highly radioactive fission products and core residues. As in other nuclear incidents the behaviour of this cloud of dangerous material would depend upon meteorological conditions, but airborne radioactivity is impossible to control and would descend on random segments of the population before any

effective warning could be given. Among those affected and heavily exposed death might ensue within seven weeks. A much larger segment of the population might be affected in a way which, much later in life, produced cancer in some form.

If such a weapon were exploded in, say, the car park outside the House of Lords, the seat of Government would be destroyed and rendered ineffective for a long period. The cost of clean-up alone, on present industrial estimates of about £1 million per gramme of plutonium, would be well over £1 billion. The cost of damage repair is inestimable but would probably be of the same order.

* Premature detonation means the thing might go off a millionth of a second too early — not while it's being built or transported...

* Safe handling of strategic materials and high explosives is widely known and described in many textbooks...

* Chemical conversion of the materials is not much riskier than heroin synthesis...and that's been going on in criminal laboratories for a long time...

94

GETTING HOLD OF THE PLUTONIUM MAY BE ONE OF THE EASIER PARTS OF THE OPERATION...THE HISTORY OF PLUTONIUM MANAGEMENT IS FULL OF UNEXPLAINED LOSSES, CARELESSNESS AND 'FIXED BOOKS'...

Accounting systems for nuclear materials can never be more than 98-99% accurate...and 1% of the annual throughput at a reprocessing plant, for example, is enough for a cartload of bombs...

Thousands of people are handling plutonium every day in the nuclear industry. There's 'statistical uncertainty' about *4 or 5 tons* of potential nuclear weapons material in the US alone — say about 1,000 bombs. Perhaps none of it has been taken...or all of it! As no one knows for sure, all threats that sound technically feasible have to be taken seriously...

WELCOME TO THE POLICE STATE...

SIPRI: 'Societies can not effectively protect the nuclear fuel cycle against sabotage short of converting to a garrison state'

In the security business it's generally agreed that you can't protect anything from people determined enough to get it...if that *were* possible there'd be no plane hijacks or bank robberies...And here again we have to take into account the human factor...

IF THE NUCLEAR POWER INDUSTRY'S INSISTENCE THAT IT CAN SAFELY GUARD THE FUEL CYCLE IS TO BE TAKEN SERIOUSLY SOME DRASTIC STEPS WILL BE NEEDED VERY SOON...

HUMAN FACTOR: DIG THIS: EVERY YEAR ABOUT 3% OF THE 120,000 PEOPLE WORKING WITH U.S. NUCLEAR WEAPONS ARE RELIEVED OF DUTY BECAUSE OF DRUG USE, MENTAL INSTABILITY OR SOME OTHER RISK.

All dissent in a Fissile Society can be viewed as a threat...so the least we can expect is an escalation of police & military surveillance...in fact what measures could not be justified in the name of NUCLEAR SECURITY?

...AND WE ARE GOING TO KEEP YOU ALL INDOORS UNTIL YOU GIVE US BACK THAT TRUCKLOAD OF PLUTONIUM!

THE WRITING'S ALREADY ON THE WALL...

Britain, we're told, is a stronghold of democratic tradition...The *Atomic Energy Authority* [*Special Constables*] *Act* of 1976, which legalized an armed nuclear police force, showed that plutonium security and democratic controls don't mix...

This little army of atomic cops *is not directly accountable to a government minister*, only to the AEA. One study has suggested that this is because it's recognized in the corridors of power that some nasty breaches of civil liberties might be needed in a plutonium emergency. In which case the government might find it convenient to plead ignorance...

In the US at least one private nuclear police force (VEPCO's) is seeking the same powers of pursuit, arrest and access to confidential files as the regular state police. And there's talk on Capitol Hill of creating a special federal force that can skip bothersome details like court orders and formal charges, and even under certain circumstances torture suspected 'nuclear terrorists'...

In many parts of the world — especially Western Europe and North America — the general public is growing more and more disturbed about nuclear power. The authorities can no longer proceed unchallenged. The public is demanding inquiries, fighting building applications and staging giant rallies and demonstrations in numerous countries...

A situation has arisen in which public participation in energy questions is seriously disrupting governments' plans for nuclear expansion...

Pushing on down the nuclear path in the face of public opposition is as good a way as any of sparking *civil disobedience*. Direct action has already begun in many countries. For example there've been...

.. SO I'M AFRAID WE HAVE TO CHANGE THE GROUNDRULES. FROM NOW ON PUBLIC INVOLVEMENT IN OUR AFFAIRS IS *ILLEGAL*!

OCCUPATIONS of proposed nuclear sites in the US, France, Britain and West Germany...

BOMB ATTACKS on nuclear installations in France, Spain, Switzerland and the US...

LEAKS of stolen confidential nuclear information in Australia...

If governments remain unswayed direct action against fuel transports and existing fuel cycle facilities also seems likely as well as interference with national electricity supplies...

AS YOU STILL WON'T ACCEPT THAT WE KNOW BEST, WE'RE CONVERTING TO A *POLICE STATE*

A TECHNOLOGY THAT SOCIETY MUST ADAPT TO...

A plutonium economy demands centralized, authoritarian rule...with fewer people wielding more power...and erosion of civil rights...starting with the right to organize labor...

Add the political and social risks to the proliferation and safety hazards of nuclear power and the picture begins to emerge of a highly oppressive technology...

WHO NEEDS NUKES?

The public?

NO THANKS...

The Scientists?

NO, THEIR TALENTS CAN BE BETTER EMPLOYED ELSEWHERE...

The Military?

YES! IT JUSTIFIES THEIR EXISTENCE!

The Engineers?

NO, FOR THE SAME REASON...

State Governments?

SOME OF THEM HAVE STAKED THEIR FUTURE ON IT. (AND OURS...)

Monopoly Capitalists?

IS IT PROFITABLE?

ONLY AS LONG AS IT IS PROFITABLE... OR IT UNDERPINS THEIR PROFITS IN OTHER SECTORS... OR PRESERVES THE POLITICAL STATUS QUO SO THAT THEY'RE FREE TO ACT...

If governments weren't propping up the nuclear industry it would soon go under. Quick profits can still be made here and there...but on the whole it's a bad investment in straight capitalist terms.

PUBLIC RESISTANCE is one of the main reasons that nuclear electricity, once hailed as a bargain, now looks a dubious economic bet. In the US, delays and safety modifications forced through by environment groups helped send *capital costs* rocketing during the 1970s...

SO FUEL PRICES ARE NOW OF SECONDARY INTEREST... AND WE BREEDERS ARE LESS ATTRACTIVE THAN EVER!

$ 100000
$ 1,000.00
$ 10000
$1.00

Capital costs of conventional power stations are increasing too, but not so fast. US analysts expect the price of nuclear electricity to catch up with the price of coal-based electricity around the mid-1980s. One of them, Barry Commoner, thinks nukes will soak up so much money that they'll lead to *capital starvation.*

IN THE END, HOW LONG THE LAME NUCLEAR DUCK IS KEPT AFLOAT MAY DEPEND ON HOW BADLY GOVERNMENTS WANT TO GET FREE OF OPEC AND ALL THOSE MILITANT COAL-MINERS....

LOOKS LIKE THE ONLY WAY WE'RE GOING TO OFF THE NUKE IS BY POLITICAL MASS ACTION

RIGHT. POPULAR MOVEMENTS LINKING **ALL** NUCLEAR OPPOSITION.

BUT WE MUST HAVE A CLEAR UNDERSTANDING OF WHAT WE'RE **AGAINST** AND WHAT WE'RE **FOR!**

PEOPLE'S POWER!

100

The sun spent 600 million years fossilizing marine life into oil and trees into coal…and at the rate we're going today they'll be used up within the next few hundred years..

NO MORE OIL…

NO MORE COAL…

GOOD-BYE GAS…

1700 1800 1900 2000

NON-RENEWABLE SOURCES OF ENERGY
like coal, oil and natural gas — and uranium, the raw material for nuclear power — have certain things in common that appeal to ruling minorities…

THAT'S GRATITUDE FOR YOU!

THEY'RE CONCENTRATED IN TERRITORY WHICH CAN BE "OWNED"…

LOTS OF MONEY IS NEEDED TO BRING THEM UP AND PROCESS THEM

A HIGH DEGREE OF TECHNICAL KNOW-HOW IS REQUIRED

£ 1,000,000

RENEWABLE SOURCES OF ENERGY
— sun, wind, water & plants — are something else…

IT'S NOT EASY TO CORNER THE MARKET IN SUNSHINE…

They're almost impossible to exhaust…they're available in many parts of many countries…and they're potentially easier and cheaper to harness…

Far too democratic for some…

…OR CENTRALIZE WIND…

SAM

IVAN

101

WHAT ARE THE SOFT ENERGY SYSTEMS?

The strength of the 'soft energy path' is its flexibility...it's a mixture of the very many techniques directly suited to people's varying energy requirements...Lovins describes them like this:

> **"** They are diverse, so that as a
> national treasury runs on many small
> tax contributions, so national energy
> supply is an aggregate of very many
> individually modest contributions,
> each designed for maximum
> effectiveness in
> particular circumstances...**"**

They're especially suitable for Third World development because they directly satisfy the basic human needs...heating...cooking... lighting...pumping...

WHEN WILL THEY BE AVAILABLE?

The picture is changing fast... a few years ago we were told the soft energy systems were of marginal interest...now the Worldwatch Institute in Washington says sunlight can supply 40% of the world's needs by the turn of the century...and a British study says we can all be living on the renewable energy sources within about 75 years...

TODAY MANY GOVERNMENTS ARE
EMBARKING ON ALTERNATIVE
ENERGY PROGRAMS...AND IT'S
THE NUCLEAR INDUSTRY THAT'S
HAVING TO CHASE FUNDS...

EXERCISE IN DOUBLE MORALITY

THE FOSSIL-FUEL BRIDGE...

To get to the stage where we can live on our energy **income** we'll have to keep drawing on our energy **savings** a while longer... half a century or more in the West...much less in the Third World...

THIS LEVELS OFF WITH IMPROVED ENERGY QUALITY AND EFFICIENCY AT POINT OF USE.

SOMETHING LIKE THIS...

Phasing out oil and natural gas means a *brief increase* in the use of coal, which is 25 times as plentiful. It has mainly been used to produce electricity and steel so far...but coal technology is undergoing a bit of a revolution these days and looks likely to replace oil in many respects...

MINING ACCIDENTS ARE GOING DOWN...

BUT WE STILL NEED A LOT MORE SAFETY MEASURES...

...AND REMOTE-HANDLING EQUIPMENT.

THE COMING THING IS FULLY-AUTOMATED MINING?

THE FISSION-FUSION FANTASY...

The nuclear industry prefers another alternative — *a fission bridge*...which means the hard technology path...which means more and more energy...more wastage...more electricity creating a need for more nukes...breeders... super-breeders...and finally...

THE GREATEST NUKE OF THEM ALL

Anti-torque structure

Plasma chamber vacuum vessel

Toroidal B-coils

Poloidal E-coils

Field shaping F-coils

The FUSION REACTOR

It slams together atoms instead of splitting them. Its raw materials are lithium and deuterium. A fusion reactor would be colossal...starting at 3,000 MWe...and make fission reactors look cheap to build. But the industry says it will be more economical to run and less dangerous...

A lasting solution of the resources problem requires careful planning of energy use, work patterns and products...social equality requires humanized production under workers' and consumers' control...a system attuned to people's real needs...not shaped by the manipulated 'demands' of passive consumers...

BUT SUBSTITUTING A PLANNED ECONOMY FOR A MARKET ECONOMY IS NOT THE ULTIMATE GOAL...ONLY A PRECONDITION FOR REAL CHANGE...

In today's market economies science and technology are largely owned and controlled by the capitalist multinationals and serve their social, political and economic goals...

In tomorrow's planned economies research and development must be strongly focused on things like public transport, socially-useful products, soft energy techniques, recycling and safe workplaces...

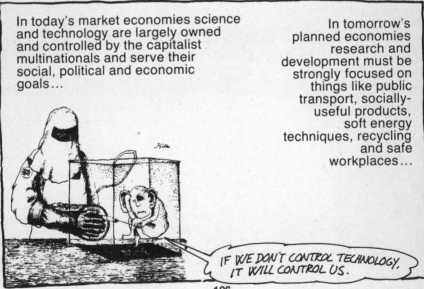

UNPLANNABLE ECONOMIES AND UNECONOMIC PLANNING...

Calls for some form of planning are now being heard from economic experts in such capitalist strongholds as Britain and the United States...without much success...

AND THAT GREY OLD KREMLIN CROWD ISN'T MUCH OF AN INSPIRATION!

The Soviet experiment went off the rails long ago...today it's more or less a parody of socialism...a centralized, authoritarian, bureaucratic and energy-hungry society riddled with nukes...

CHOOSING THE SOFT ENERGY PATH AND RECOGNIZING THE PHYSICAL AND BIOLOGICAL CONSTRAINTS ON HUMAN ACTIVITY DOESN'T MEAN FREEZING IN THE DARK...

It means choosing a *reasonable* material standard and a *reasonable* level of energy use instead of growing inequality and alienation...it means placing greater value on things that *count* instead of things that are merely *countable*...

We're all basically creative... It's only when we're not given responsibility or the tools to control our own lives that we become passive, violent, neurotic...

Active people who like to experiment and solve problems are not easy to oppress... they develop the self-confidence that is the lifeblood of real democracy...they don't need to be mobilized, led, ruled or watched over...quite literally, they're capable of taking matters into their own hands...

108

AND FINALLY:
CHAPTER 6.

In which the environmentalists start the ball rolling, the workers tackle them, both revise their ideas and just about everyone joins the anti-nuclear struggle...

LIBERAL · TRADE UNIONIST · ECO-LOGIST · RADICAL · CONSERVATION-IST · CONCERNED CITIZEN · CRITICAL SCIENTIST

HELL NO, WE WON'T GLOW...

Opponents of nuclear power come in many shapes and sizes...

They're more or less agreed on the dangers of nuclear technology and on the advantages of the soft energy path...but they adopt widely differing approaches to the problem...with varying degrees of success...

AS A SPECIES, ANTI-NUKERS FIRST CROPPED UP IN THE UNITED STATES IN THE EARLY 1960s. BY THE LATE 1970s THEY HAD SPREAD EN MASSE TO WESTERN EUROPE, JAPAN AND AUSTRALIA.

In the beginning, resistance mainly took the form of lobbying government or challenging the nuclear builders at license hearings, public inquiries and in court...

Today, the emphasis is shifting more and more to direct, non-lobbying methods and to building a mass movement against nuclear power...sometimes linking up across national boundaries...

PSST... OFF THE NUKE!

OFF THE NUKE!

RADIATION IGNORES BORDERS TOO...

110

Once, it was little more than a technical discussion between experts for the benefit of the politicians...

Now, the energy debate has become a matter for the general public, not just the decision-makers...and the citizenry isn't taking too kindly to being force-fed with nukes...

ALMELO,
NETHERLANDS:
30,000 PROTEST...

MALVILLE,
FRANCE:
50,000 PROTEST...

MELBOURNE,
AUSTRALIA:
20,000 PROTEST...

SEABROOK,
UNITED STATES:
1,500 ARRESTED!

MONTALTO, ITALY:
10,000 PROTEST...

BILBAO, SPAIN:
150,000 PROTEST...

BROKDORF,
WEST GERMANY:
40,000 PROTEST...

LONDON,
ENGLAND:
10,000 PROTEST...

Not to mention Sam Lovejoy...

❝❝ On George Washington's Birthday, 1974, Samuel Holden Lovejoy toppled a 500-foot-tall weather tower in Montague, Massachussets. The tower had been erected by the local utility company as part of its project to construct one of the largest nuclear power plants ever planned. Leaving 349 feet of twisted wreckage behind, Lovejoy hitched a ride to the police station, where he turned himself in and submitted a four-page written statement decrying the dangers of nuclear power. Six months later, Lovejoy defended his act of civil disobedience in court as «self-defense». He was ultimately acquitted. ❞❞

The anti-nuclear power movement is growing militant with the realization that you can't de-rail the Nuclear Express by gentle persuasion...

111

THE ECOLOGISTS POINTED THE WAY...

Environment groups launched the anti-nuke movement and have long been its vanguard...they were the first to spell out the dangers of a hard energy future...to organize resistance...and to formulate alternatives...

Adding an ecological dimension to the class struggle, radical environmentalists offered a fresh solution to the ravages of capitalism...but they also annoyed many dogged Marxists who kept insisting that technology was neutral...

BUT SOME ECOLOGISTS LOST THEIR WAY...

112

OTHERS GOT CARRIED AWAY...

...AND MANY MISSED THE POINT...

SO IT WASN'T STRANGE THAT MUCH OF THE LABOR MOVEMENT DISMISSED THE ENVIRONMENT MOVEMENT AS AN ASSEMBLY OF MIDDLE-CLASS CRAZIES OUT OF TOUCH WITH THE FACTS OF WORKING LIFE...

WORKERS IN SOME COUNTRIES STARTED LISTENING WHEN
ENVIRONMENTALISTS STARTED TALKING THEIR LANGUAGE...

DEMANDS FOR ALTERNATIVE PRODUCTS AND MANUFACTURING METHODS BRING THE LABOR MOVEMENT TO THE FOREFRONT OF THE ENVIRONMENT STRUGGLE...

Alternative production is not much good without alternative consumption! Socially useful products are useless if they don't sell!

The market economy is insensitive to social needs...but this must be *seen* to be the case! Groups outside the factory — environmentalists, consumers, unionists in the service sector — can help by bringing pressure on public institutions *to order the alternative goods to meet obvious needs...*

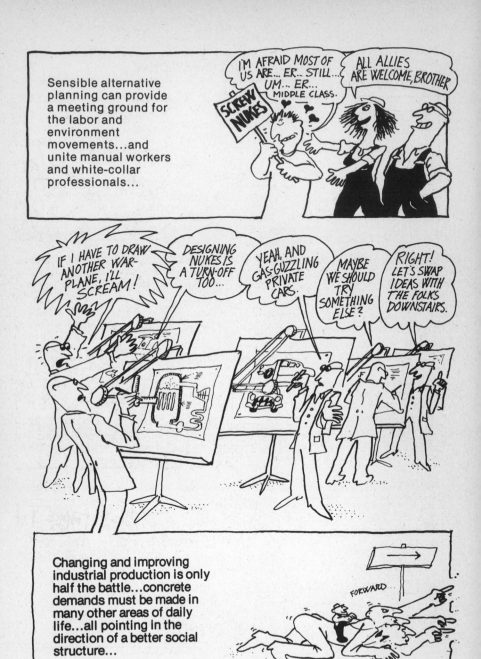

116

TOWARDS A TOTAL PERSPECTIVE...

In the Public Services...

In housing areas...

At the shops...

In public transport...

...AND A CO-ORDINATED DEMOCRATIC MOVEMENT!

To quote those wise old ideological fathers of socialism Marx & Engels...

A REAL MOVEMENT...

... BEATS TWELVE PROGRAMS!

Marx wasn't exactly an eco-freak. He did criticize capitalist agriculture and the dehumanizing effect of separating people from nature...but he also tended to view capitalist technology as progressive. After him, Lenin, Trotsky, Stalin and other Marxists thought you could just take it over and paint it red.

IF I'D KNOWN ABOUT FAST BREEDERS I MIGHT HAVE HAD SECOND THOUGHTS.

ANYWAY, MARXISM IS NOT JUST A THEORY... IT'S A WEAPON TO BE CONSTANTLY SHARPENED AND USED IN NEW WAYS.

But Marx & Engels were good at driving home one point — that all efforts to bring about fundamental social change must be mass-based if they're to succeed!

SO...

118

TALK TO FRIENDS…NEIGHBORS…RELATIVES…FELLOW-
WORKERS…TAKE PART IN ACTIONS…JOIN GROUPS…GO TO
MEETINGS…GET ACTIVE IN YOUR UNION…OR YOUR LOCAL
COMMUNITY…OR BOTH…KEEP INFORMED…USE YOUR
IMAGINATION…

READING AND DISCUSSING IS ONE THING

BUT IT'S MORE IMPORTANT TO ACT… TO GET **INVOLVED.**

**BETTER ACTIVE TODAY
THAN RADIOACTIVE TOMORROW**

**IN THE END
IT'S UP TO YOU**

APPENDIX.

Alternative Energy Systems

The Sun...

Techniques harnessing the energy of the sun are being developed
swiftly and widely. More than a century ago solar energy was used in
Chile to desalinate water and in the US to power printing presses.
Today it is in use in many fields and new applications are constantly
being discovered.

FLAT-PLATE SOLAR
COLLECTORS for
domestic heating are
the commonest
hardware. They have
long been in use in
hot countries like
Israel, South Africa
and Australia. Now
they are in great
demand in the US,
where sales are
trebling annually.

*Heating water in a West German
swimming pool...and a Welsh cottage*

Collectors are also finding a market
in cloudier regions like Scandinavia
and Britain as a source of hot water
in summer and low-temperature
heating in winter. The heat is usually
stored in water or rockbeds. Solar
collectors can be applied to both
large and small premises,
'back-fitted' to existing buildings or,
more simply, fitted into the structure
from the start. They are most
economically used by groups of
households sharing storage.

SOLAR FOCUSING COLLECTORS reflect and concentrate the sun's rays on one spot. They range from gigantic high-temperature furnaces capable of melting metals to do-it-yourself cookers for boiling water and baking bread, etc.

French solar furnace...Third World grill made of bamboo, clay and foil...and a bank of solar [photo-voltaic] cells for the rich

SOLAR CELLS turn sunshine into electricity. They are still extremely expensive but a major cost breakthrough is said to be imminent.

In the US, government and private capital is being poured into solar cell technology, which lends itself to monopoly. The same can be said of SOLAR POWER TOWERS, electric energy systems now being built in Sicily, France, Spain and the US. Ground mirrors provide the heat to turn a steam turbine in the tower.

122

The Wind...

An endless source of energy with no pollutive side-effects, the wind has been put to good use since 2000 BC when the Chinese and Persians ground corn with it. Current **WIND POWER** systems can heat, pump, compress air and generate electricity. Designs range from massive turbines on 50-meter towers geared to electricity grid supply...

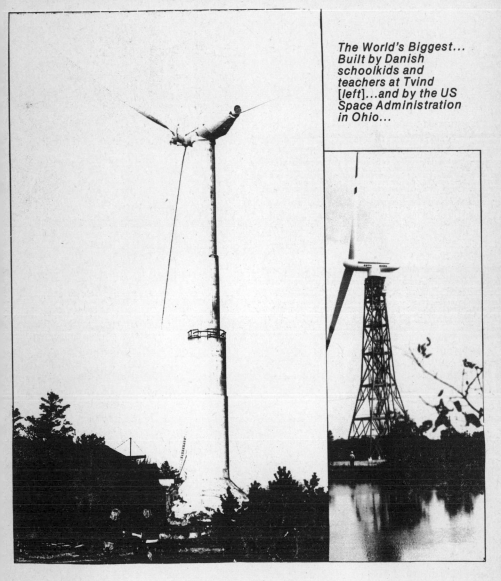

*The World's Biggest...
Built by Danish
schoolkids and
teachers at Tvind
[left]...and by the US
Space Administration
in Ohio...*

...down to the spinning oil drum variety for less ambitious ventures. The Greek island of Crete has over 100,000 small windmills pumping water for irrigation.

The Savonius Rotor...heavy but cheap and simple...and another vertical-axle wind generator, the high-speed Darrius which the Danes and Canadians think may produce 1.5 MWe...

Wind energy is highly versatile and an aid to decentralization. It can be used in many ways at the point of use, reducing the need for long, expensive and wasteful powerlines. Battery storage is most common but there's talk of an 'inverter' that switches over to the grid when there's no breeze around...and also works the other way, turning your electric meter backwards!

British studies suggest that unit-for-unit windpower's total cost is one-third that of nuclear power...and the World Meteorological Office says 20 million MWe is blowing about...

But the wind does not always blow and the sun does not always shine...so a mixture of the two (above) is useful. Good insulation lessens the need for both by reducing a building's heat requirements.

124

The Water...

Harnessing the mechanical energy in falling water is the oldest technique in the book. The modern successor to yesterday's creaking water wheels is the streamlined turbine generating electricity, often in huge dam projects. These frequently damage the environment and disrupt local cultures. But it is no longer accepted that only large volumes of water can be economical and there is growing interest in medium-sized

HYDRO-ELECTRIC POWER PLANTS,

especially in Finland and the Soviet Union , while the plunging cost of small water turbines is making 'micro-stations' look attractive once more. Many countries have idle water wheels that can be revived to make electricity — there are 20,000 in England alone. China has built 60,000 in 15 years to supply 20% of her electricity needs. Most of the world's untapped water power is in the developing countries, especially in Africa, which has 20% of the global potential but generates a mere 2%.

Water power is the most reliable of the renewable energy sources. Except in extreme droughts it flows non-stop, strongest in winter when demand is greatest. However, most waterways these days are 'owned'...

125

TIDAL POWER SYSTEMS exploit the ebb and flow of the tides so are not dependent on climate. A small scale station is operating in France, a larger one is being built in Canada's Bay of Fundy and a giant tidal power barrage across Britain's River Severn is under consideration.

La Rance has been providing 240 MWe for 10 years...and two British wave-power designs with built-in turbines...

WAVE POWER techniques harnessing the energy of the ocean waves are under development mainly in Britain, Japan and Scandinavia. The Norwegians estimate that a 150-kilometer string of wave power units could supply 70,000 million kilowatt hours of electricity — the nation's current annual consumption.

OCEAN THERMAL POWER UNITS exploiting the difference in temperature between the sea depths and the surface are soaking up a lot of US government funds in anticipation of a mid-1980s market...

Wave contouring raft.

Salter's ducks.

126

The Earth...

Plants solved the energy crisis 3 billion years ago when they learned to create themselves out of earth, air and water. Now they're back in favo as 'bio-mass crops' or **BIO-FUELS**.

Apart from direct wood-burning, methods exist to tap the energy stored in plants, trees and organic waste by converting them into methane and other liquid and gaseous fuels for the transport sector. Special 'energy plantations' of fast-growing trees like alder are being cultivated while ocean farming of giant seaweeds and other marine plants is also under way. Large-scale organic conversion presents some environmental problems like soil depletion and requires careful integration in forestry and agriculture.

Small-scale production of methane is a well-tested technology among Chinese and Indian farmers (right) whose air-tight digesters are mostly filled with cow-dung and human sewage. It is estimated that animal waste and crop residues on US farms could supply all the power needs of that country's enormous agricultural industry.

RECYCLING metals, glass, wood, paper, cloth, oil, batteries and even some plastics is another energy-saver...'waste' is a raw material in the wrong place at the wrong time!

GEO-THERMAL ENERGY, drawing on the heat deep in the Earth's crust, is easiest to plumb in volcanic regions but will become available elsewhere with better drilling techniques. Another centralized, high-technology system for the energy corporations and their state backers.

HEAT PUMPS, drawing on the Earth's warmth or the air, are already in widespread use. Many are powered from the mains. New designs link heat pumps to wind and solar energy systems.

Recycling scrap metal...geo-thermally heating a Paris block of flats...and pump-heating a row of 12 Colorado apartments...

128

URANIUM PRODUCTION (1977)
(TONS)

Argentina	130
Australia	400
Canada	6,100
France	2,200
Niger	1,609
Portugal	85
South Africa	6,700
Spain	191
US	11,200

Total (14 non-socialist countries)
28,617 tons

ACTUAL AND ESTIMATED URANIUM REQUIREMENTS

Source: Int. Symposium on Uranium Supply and Demand, London July 1978

	1977	1980	1985	1990
Non socialist world. (rounded)	24,400	40,500	76,000	111,000

ORGANIZATIONS AGAINST NUCLEAR POWER

Australia

Movement Against Uranium Mining (MAUM), 277 Brunswick St, Fitzroy, Victoria 3065.

Community Research Action Centre, Monash University Union, Clayton, Victoria 3168.

Friends of the Earth, 232 Castlereagh St, Sydney, New South Wales 2001.

Austria

Initiative Österreichischer Atomkraftwerksgegner (IOAG), Postfach 138, 1071 Vienna.

Belgium

Amis de la Terre, Rue E. Walschaert 19, 1060 Bruxelles.

Verenigde Aktiegroepen voor Kernstop (VAKS), Consciencestraat 46, 2000 Antwerpen.

Britain

Socialist Environment & Resources Association (SERA), 9 Poland St, London W1V 3DG.

Scottish Campaign to Resist the Atomic Menace (SCRAM), 2a Ainslyplace, Edinburgh 3.

Friends of the Earth, 9 Poland St, London W1V 3DG.

Nuclear Information Network (NIN), 29 Great James St, London WC1.

London Greenpeace, 6 Endsleigh St, London WC1.

Colonialism & Indigenous Minorities Research Action (CIMRA), 70 Durham Rd, London N7.

Canada

Canadian Coalition for Nuclear Responsibility (CCNR), 2010 Mackay St, Montreal, Quebec H3G 2J1.

Saskatoon Environmental Society, PO Box 1372, Saskatoon, Saskatchewan S7K 3N9.

Maritime Energy Coalition, 77 rue Germaine, St. John, New Brunswick

Société pour Vaincre la Pollution, P.O. Box 65, Place d'Armes, Montréal, Quebec

Energy Probe, 43 Queens Park Crescent East, Toronto, Ontario

Ontario Coalition for Nuclear Responsibility, Box 872, Waterloo, Ontario

Greenpeace, 2108 W. 4th Avenue, Vancouver, British Columbia

Denmark

Organisationen til Oplysning om Atomkraft (OOA), Skindergade 26, DK-1159 Copenhagen K.

Finland

Alternative till Kärnkraft, PB 143, 00201 Helsinki 20.

France

Coordination Antinucléaire Région Parisienne, C/o Yves Behar, 3 rue Félix Ziem, 75018 Paris.

Comité contre la pollution dans la Hague, BP 156, 50104 Cherbourg Cédex.

Comité contre le Super-Phénix (Malville), C/o Marie Dubost, Iselet, Morestel.

Amis de la Terre, 117 Av. de Choisy, 75013 Paris.

Comité de Sauvegarde de Fessenheim et de la plaine du Rhin (CSFR), C/o J.J. Rettig, École de Champenay, 67420 Saales.

West Germany

Bundersverband Bürgerinitiatieve Umweltschutz (BBU), Schiffkopfweg 31a, 75 Karlsruhe 21.

Burgerinitiatieve Umweltschutz Unterelbe (BUU), Schlüterstr. 4, 2000 Hamburg 13.

Anti-AKW-Laden, Lutterothstr. 33, 2000 Hamburg 19.

Bürgeraktion Küste (BAK), Bückerburgerstr. 50, Bremen 2800.

Gruppe Internationalismus in der Bremer Bürgerinitiative gegen Atomanlagen, Fedelhören 14, 28 Bremen 1.

Ireland

C/o John Carroll, Irish TGWU, Liberty Hall, Dublin 1.

Japan

Jishu Koza, Masafumi Takubo, B. Kaikan, Nishi Okubo 2-350, Shinjukuku, Tokyo.

Luxemburg

Bürgerinitiative Museldal, C/o Jemp Weydert, 51 rue demy Schlechter, Luxemburg.

Netherlands

Landelijk Energie Kommitee (LEK), 2e Weteringplantsoen 9, Amsterdam.

New Zealand

Friends of the Earth, Box 39-065, Auckland West.

Norway

Aksjon mot Atomkraft, PB 8395, Hammersborg, N. Oslo 1.

Spain

Comision de Defensa de una Costa Vasca no Nuclear, Avenida Basagoiti 28, Algorta.

Comité Antinuclear de Catalunya (CANC), Bruc 26 2on., Barcelona 10.

Sweden

Folkkampanjen mot Atomkraft, Tjärhovsgatan 44, 11629 Stockholm.

Miljöförbundet, Box 2129, 750 02 Uppsala.

Switzerland

National Koordination der Schweizer Anti-AKW-Organisationen, C/o André Froidevaux, Burgunderstr. 4, 4051 Basel.

Schweizerische Energiestiftung (SES), Auf der Mauer 6, 8001 Zurich.

United States — National

Citizens' Outreach for a No Nukes Education of Congress Today (CONNECT), 4213 Kennedy Street, Hyattsville, MD 20781, 301 699-9438

Nuclear Information and Resource Service, 1536 16th Street NW, Washington DC 20036, 202 347-8317

Mobilization for Survival, 1213 Race Street, Philadelphia, PA 19107, 215 563-1512

Natural Resources Defense Council, 917 15th Street, Washington DC 20005, 202 737-5000

Friends of the Earth, 124 Spear Street, San Francisco, CA 94105, 415 495-5770

Environmental Policy Center, 317 Pennsylvania Avenue SE, Washington DC 20003, 202 547-6500

Campaign on Runaway Reactors, 225 4th Street NE, Washington DC 20002, 202 547-1656

Critical Mass, P.O. Box 1538, Washington DC, 202 546-4790

National Public Interest Research Group, 1329 E Street NW, Suite 1127, Washington, DC 20004, 202 347-3811

Nuclear Cargo Task Force, P.O. Box 2234, High Point, NC 27261, 919 882-0109

East

Friends of the Earth, 72 Jane Street, New York, NY 10014, 212 675-5911

Union of Concerned Scientists, 1208 Massachusetts Avenue, Cambridge, MA 02138, 617 547-5552

American Friends Service Committee, 2161 Massachusetts Avenue, Cambridge, MA 02140, 617 661-6130

Clamshell Alliance, 62 Congress Street, Portsmouth, NH 03801, 603 436-5414

Keystone Alliance, 1006 S. 46th Street, Philadelphia, PA 19143, 215 387-5254

Midwest

Bailly Alliance, P.O. Box 847, Chesterton, IN 46304

Paddlewheel Alliance, P.O. Box 344, Bloomington, IN 47401

Citizens United for Responsible Energy, 1342 30th Street, Des Moines, IO

Citizens' Action For Safe Energy, Inc., P.O. Box 924, Claremore, OK 74107, 918 341-3494

South and Southwest

Lone Star Alliance, 2521 Enfield, Austin, TX 78703

People's Action for Safe Energy, 401 Watson, Fayetteville, AR 72701

Palmetto Alliance, 18 Bluff Road, Columbia, SC 29201, 803 771-9999

Potomac Alliance, P.O. Box 9036, Washington DC 20005

Catfish Alliance, Tallahassee, FL 36078

West

Northern California Alliance for Survival, 1360 Howard Street, San Francisco, CA 94103, 415 626-1465

Abalone Alliance Clearinghouse, 452 Higuera Street, San Luis Obispo, CA 93401, 805 543-6614

Southern California Alliance for Survival, 5539 W. Pico Boulevard, Los Angeles, CA 90019, 213 937-0240

Committee for Nuclear Responsibility, Box 11207, San Francisco, CA 94101

Headwaters Alliance, 430 North Higgins, Missoula, MT, 406 721-1453

Rocky Flats Truth Force, 972 Pleasant Street, Boulder, CO

EARS, 2239 E. Colfax, Denver, CO 80206, 303 321-1645

S.W. Research and Information, P.O. Box 4524, Albuquerque, NM 87106

Crabshell Alliance, Box 7185, Olympia, WA 98507, 206 866-6191, and 1114 34th Avenue, Seattle, WA 98122, 206 323-2880

Trojan Decommissioning Alliance, 215 S.E. 9th, Portland, OR 97214, 503 231-0014, and 348 W. 8th Street, Eugene, OR 97401, 503 687-2557

General

World Information Service on Energy (WISE), 2e Weteringplantsoen 9, Amsterdam, Netherlands.

International Mobilisation for Survival, C/o ICDP, 6 Endsleigh St, London WC1, Britain.

TIPS FOR FURTHER READING

Books

Berger, J. **Nuclear Power: The Unviable Option.** Dell, 1977.
Bookchin, M. **Post-Scarcity Anarchism.** Ramparts, 1971.
Commoner, Barry. **The Poverty of Power.** Bantam, 1977.
Elliot, D., ed. **The Politics of Nuclear Power.** Pluto Press, 1978 (London).
Fuller, John G. **We Almost Lost Detroit.** Ballantine, 1976.
Harper, P., and Boyle, G., eds. **Radical Technology.** Pantheon, 1976.
Hayes, Denis. **Rays of Hope: The Transition to a Post-Petroleum World.**
Norton, 1977.
Illich, Ivan. **Tools for Conviviality.** Harper & Row, 1974.
Lanoue, Ron. **Nuclear Plants: The More They Build, The More You Pay.**
Critical Mass, 1976.
Lovins, Amory. **Soft Energy Paths.** Penguin, 1977.
Nader, Ralph, and Abbotts, John. **The Menace of Atomic Energy.**
Norton, 1977.
Patterson, W. **The Fissile Society.** Earth Resources, 1977.
———. **Nuclear Power.** Penguin, 1976.
Roberts and Medvedev. **The Hazards of Nuclear Power.** Spokesman,
1978 (London).
Smith, D., ed. **The Big Red Nuclear Diary.** Pluto Press, 1978.
Union of Concerned Scientists. **The Nuclear Fuel Cycle: A Survey of the
Public Health, Environmental and National Security Effects of Nuclear
Power.** UCS, Boston.
———. **The Risks of Nuclear Power Reactors.** UCS, Boston, 1977.

Periodicals

Undercurrents, 27 Clerkenwell Clo, London EC1R 0AT, Britain.
New Scientist, 128 Long Acre, London WC2E 9OH.
Science for People, 9 Poland St, London W1.
Science for the People, 9 Walden St, Jamaica Plain, MA 02130, USA.
Bulletin of the Atomic Scientists, 1020-24 East 58th St, Chicago, Ill 60637.
Environment, 438 N. Skinker Boulevard, St Louis, Mo. 63130.
Spark, CSRE, 475 Riverside Drive, New York, NY 10027.
Radical Ecologist, PO Box 87, Carlton South, Victoria 3053, Australia.
Chain Reaction, from FOE, 232 Castlereagh St, Sydney, NSW 2001, Australia.
Agenor, 13 Rue Hobbema, 1040 Brussels, Belgium.
La Gueule Ouverte, BP 26, 71800 La Clayette, France.
Survivre et Vivre, 6 Rue Chappe, 75018, Paris, France.

About the Author and Illustrator

Stephen Croall is a British writer and journalist currently living on a farm in Sweden. A former Reuter correspondent in Stockholm and London, he has also worked as a musician, a shoemaker, and a day nursery apprentice.

Kaianders is an American designer and cartoonist also living in Sweden. A qualified architect, devoted indoor gardener, and recycler of tandems, he has provided illustrations for hundreds of album covers, magazines, and books in Sweden.